Climate fluctuations can trigger events that lead to mass migration, hunger and famine. Rather than attributing the blame to nature, the authors look at the underlying causes of social vulnerability, such as social processes and organization. Past and present susceptibility to destitution, hunger, and famine in the face of climate variability can teach us about the potential future consequences of climate change. By understanding why individuals, households, nations and regions are vulnerable, and how they have buffered themselves against climatic and environmental fluctuations, present and future vulnerability can be redressed. Through case studies from around the globe, the authors explore past experiences with climate variability, as well as the likely effects of, and the possible policy responses to, the types of climatic events that global warming might bring.

Climate Variability, Climate Change and Social Vulnerability in the Semi-arid Tropics

Climate Variability, Climate Change and Social Vulnerability in the Semi-arid Tropics

Edited by

Jesse C. Ribot

MacArthur Fellow,
Center for Population and Development Studies,
Harvard University, Cambridge, Massachusetts, USA

Antonio Rocha Magalhães

Executive Secretary,
Secretariat of Planning,
The Presidency of the Federated Republic of Brazil, Brasilia, Brazil

Stahis S. Panagides

Vice President,
Esquel Group Foundation, Bethesda, Maryland, USA

CAMBRIDGE
UNIVERSITY PRESS

CAMBRIDGE UNIVERSITY PRESS
Cambridge, New York, Melbourne, Madrid, Cape Town, Singapore, São Paulo

Cambridge University Press
The Edinburgh Building, Cambridge CB2 2RU, UK

Published in the United States of America by Cambridge University Press, New York

www.cambridge.org
Information on this title: www.cambridge.org/9780521480741

First published 1996
This digitally printed first paperback version 2005

A catalogue record for this publication is available from the British Library

Library of Congress Cataloguing in Publication data

Climate variability, climate change, and social vulnerability in the
semi-arid tropics / edited by Jesse C. Ribot, Antonio Rocha
Magalhães, Stahis S. Panagides.
    p.   cm.
ISBN 0 521 48074 4
1. Climatic changes–Social aspects–Tropics–Congresses. 2. Arid
regions–Climate–Social aspects–Congresses. 3. Sustainable
development–Tropics–Congresses. I. Ribot, Jesse Craig. II. Magalhães,
Antonio Rocha. III. Panagides, Stahis Solomon, 1937– .
QC981.8.C5C62 1996
304.2′5′0913–dc20 95-1625 CIP

ISBN-13 978-0-521-48074-1 hardback
ISBN-10 0-521-48074-4 hardback

ISBN-13 978-0-521-01947-7 paperback
ISBN-10 0-521-01947-8 paperback

Contents

Contributors

Tania Bacelar de Araújo	Economist, Social Research Institute of the Joaquim Nabuco Foundation, Recife, Brazil
Jan Bitoun	Professor of Geography, Department of Geographical Science, Federal University of Pernambuco, Recife, Brazil
Michael H. Glantz	Senior Scientist and Director, Environmental and Societal Impacts Group, National Center for Atmospheric Research (NCAR), Boulder, Colorado, USA
Leonardo Guimarães Neto	Economist, Social Research Institute of the Joaquim Nabuco Foundation, Recife, Brazil
R. Les Heathcote	Professor of Geography, Flinders University, Adelaide, South Australia
Diana Liverman	Associate Professor of Geography and Earth Systems Science, Pennsylvania State University, University Park, USA
Adil Najam	Doctoral Candidate, Department of Urban Studies and Planning, Massachusetts Institute of Technology, Massachusetts, USA
Karen O'Brien	PhD Candidate in Geography, Pennsylvania State University, University Park, Pennsylvania, USA
Jesse C. Ribot	MacArthur Fellow, Center for Population and Development Studies, Harvard University, Cambridge, Massachusetts, USA
Fredrick J. Wang'ati	Technical Coordinator, United Nations Environment Program, Desertification Control Program Activity Center, National Land Degradation and Assessment and Mapping in Kenya, Nairobi, Kenya
Gabrielle Watson	Consultant, World Bank, Washington, DC, USA
Donald A. Wilhite	Professor of Agricultural Climatology, Director of International Drought Information Center, Department of Agricultural Meteorology, University of Nebraska, Lincoln, Nebraska, USA
Zing-ci Zhao	Professor, Chinese Academy of Meteorological Sciences, Beijing, People's Republic of China

Foreword

The threat of climate change has united scholars, practitioners, policy-makers and many publics to challenge the foundations of wasteful economic systems and, now over two years ago, to forge an unprecedented treaty: the Framework Convention on Climate Change. What is the contribution of the academic community to meet this challenge? Obviously, understanding of climate systems and their interactions with the biogeosphere brought the issue to the world's attention, garners the majority of funding, and, indeed, continues as an urgent need.

Yet, an equal contribution, in my view, is required from social scientists. The many disciplines are replete with frameworks (for example, political ecology, sustainable development, and risk assessment), concepts and methods (from structuration to contingent valuation and participatory rural appraisal), and prescriptions for action (such as community self-help and empowerment embodied in the slogan 'Think global, act local'). Across this diverse landscape, three foci should be prominent.

First, vulnerable populations and social equity must be firmly embedded in the science and politics of global change. Impact assessments, economic evaluations and international negotiations must be cognizant of the great disparities in livelihoods. Those who have contributed the most to climate change are able to bear the consequences more readily than those whose livelihoods are under threat at present and can least afford to either prevent climate change or survive potentially adverse consequences.

Second, the nuances of local vulnerability and capability must be clearly understood. In the near term, research on the human ecology of production, access to markets, and the politics of empowerment must be conducted at a local level. Regional and global patterns will only emerge from careful, consistent case studies of individuals, households and communities.

Third, local studies must be placed in the context of global change. Such transitions as from self-provisioning to mixed economies, from authoritarian to democratic government,

from individual to collective responsibility imply a spatial connectedness that spans the world. At the same time, the dynamics of present resource use, social conditions, economic systems and political societies must be projected forward to match the time scale of climate change. Forecasting the evolution of vulnerable groups and regions remains one of the most perplexing issues in understanding the potential impacts of global change. For example, will famine continue to plague Africa in the next century? Will famine continue despite international agreements to end famine by the end of the 1990s and despite projections of economic growth such that developing countries in 2100 will be as rich then as the OECD countries are now?

This volume brings fresh insight to these issues. It firmly focuses attention on the conjuncture of the driving forces of global change, resource use in the climatically marginal semi-arid regions of Africa, Asia and Latin America, and the social geography of vulnerability. The commitment of the authors and editors provides a contribution to the literature on sustainable development and natural disasters. Furthermore, it looks beyond current approaches of climate change impact assessment toward fresh ways to identify vulnerable places, link the local and the global, and address the urgent needs of present vulnerable populations with the long-term need to enhance their capacity to respond to the effects of climate change. This latter objective should provide valuable insight for efforts to draft and adopt a protocol on adaptation for the Framework Convention on Climate Change.

It was a pleasure to have contributed to the International Conference on Impacts of Climatic Variations and Sustainable Development in Semi-Arid Regions in Fortaleza in 1992. It is an equal pleasure to recommend this book to a wide audience.

Thomas E. Downing
Programme Leader for Climate Impacts and Responses
Environmental Change Unit
Oxford

Preface

It is with great satisfaction that we present this edited volume. It draws from the rich and timely presentations at the International Conference on Impacts of Climatic Variations and Sustainable Development in Semi-Arid Regions (ICID) held in Fortaleza, State of Ceará, Brazil, 27 January to 1 February 1992. ICID was a concerted effort on the part of social scientists, climatologists, policy analysts and policy makers to examine, find solutions for and bring attention to the common problems faced by the peoples of semi-arid lands – the most profound of which are associated with climatic phenomena. Conference participants were asked to examine the past consequences of and responses to climate variability, and given these past experiences, to reflect on the likely effects of and possible proactive policy responses to the types of climatic events that global warming might bring. Nations and peoples of the semi-arid regions of the world have had long histories of planning for, coping with, rebuilding after, and responding to variations inherent in their climates. The papers in this volume recount some of these histories and reflect on the future of the semi-arid lands of the least-developed countries.

ICID preceded the United Nations Conference on Environment and Development (UNCED), held in Rio de Janeiro in June 1992, in order to give voice at UNCED to the plight and needs of peoples of these semi-arid lands. The semi-arid regions of the world contain a large portion of the least-developed countries on earth. It is in these poorer semi-arid lands – the semi-arid tropics – where populations eke out their livings squeezed against a fluctuating and at times declining natural resource base. The ultimate need for development – throught entitlement and empowerment – to moderate the relation between people and their environment is manifest here. UNCED promised to address the problems of this very relationship between people and the land in the underdeveloped world. But, in the international arena of UNCED, world attention was focused on global-level or macro aspects of issues such as climate change, biodiversity and deforestation. Concerns about the living relation between local peoples and their natural resource base were overshadowed by these global concerns. The most underde-

veloped regions, where the majority's livelihoods are caught between the limits of environment and development, were marginalized in the very context in which their concerns should have been center stage. ICID helped to bring these concerns back into the debate by introducing the conference declaration, the Declaration of Fortaleza (this volume), into the last preparatory session leading up to UNCED. But the real results of this effort will unfold in the years to come as Agenda 21 (UNCED's global plan of action) and other international environment agreements and institutions, such as the recent convention on desertification, take shape as policies on the ground. It is our hope that this volume can contribute to the debates that shape such policies.

The authors and editors of this volume are grateful to the sponsors, organizers and supporters of ICID. The Conference was convened by the government of Ceará and organized by Esquel Group Foundation Brazil (Fundação Grupo Esquel Brasil). Initial and generous support for this event was rendered by the Bank of the Northeast of Brazil and the Ceará Federation of Industries. Contributions were received from other Brazilian institutions, among them the National Science Council (CNPq), the Brazilian Agricultural Research Corporation (EMBRAPA) through its Research Center on the Semi-Arid Tropics (CPATSA), and the J. Macedo Group. An indication of the increasing awareness by the international community regarding the importance of semi-arid lands was the assistance received from sources outside Brazil. The Inter-American Development Bank (IDB), the Government of the Netherlands and the John D. and Catherine T. MacArthur Foundation were our most generous external financial supporters. Additional international support was received from the United Nations Environment Program (UNEP), the French Institute for Ultramarine Research (ORSTOM), and the World Bank. From its very beginning, ICID has received encouragement and guidance from Mr Maurice Strong, Secretary General of UNCED.

Special appreciation is due to the former Governor of the State of Ceará, Mr Ciro Gomes, and to the present Governor, Mr Tasso Jereissati. It was their belief that science, international awareness, and political commitment must move together that made this event possible. Special thanks are due to Esquel Group Foundation for its continuous support and leadership role in ICID. Esquel Group Foundation USA, and the Esquel Group Foundation Brazil, are members of the eight country Grupo Esquel network that assisted ICID throughout its lengthy and laborious preparatory process. The Massachusetts Institute of Technology (MIT), through its Department of Urban Studies and Planning, provided us invaluable guidance and professional support.

The cooperation among the teams in Fortaleza, Brasilia, Cambridge and Washington in preparing this volume, as well as in convening and organizing ICID, is a vivid example of how civil society, the scientific community and government can cooperate across borders and disciplines to focus attention on problems that influence the destiny of us all.

A. R. M.
S. S. P.
J. C. R.

Abbreviations

BCE	Before the common era, referring to the time often called BC
BP	Before present, present is 1950
CFC	Chlorofluorocarbon: a gas known to destroy the earth's ozone layer
CILSS	Comité Inter-état pour la lutte Contre la Sacheresse au Sahel (Inter-State Committee for the Fight Against Drought in the Sahel)
ENSO	El Niño Southern Oscillation: a weather pattern that is an important determinant of global climatic patterns
FAO	Food and Agriculture Organization of the United Nations
GCM	General Circulation Model: models used by climatologists to forecast weather patterns
GHG	Greenhouse gases: gases that contribute to global warming by trapping the sun's heat in the earth's atmosphere
ICID	International Conference on Impacts of Climatic Variations and Sustainable Development in Semi-Arid Regions
ICRISAT	International Crops Research Institute for the Semi-arid Tropics
IGADD	Inter-Governmental Authority on Drought and Development, East Africa
IPCC	Inter-Governmental Panel on Climate Change of the United Nations
MNCs	Multi-National Corporations
NEB	Northeast Brazil
P/ETP	Precipitation to potential evapotranspiration ratio: a measure of aridity
PDSI	Palmer drought severity index
SDS	Sustainable development strategy
SUDENE	Superintendencia do Desenvolvimento do Nordeste (Northeast Brazil Development Agency)
UCAR	University Corporation for Atmospheric Research, USA
UNCDF	United Nations Capital Development Fund
UNCED	United Nations Conference on Environment and Development (June 1992, Rio de Janeiro)
UNDP	United Nations Development Program
UNEP	United Nations Environment Program
WCED	World Commission on Environment and Development (Brundtland Commission)
WFP	World Food Program of the United Nations
WMO	World Meteorological Organization

Introduction. Climate Variability, Climate Change and Vulnerability: Moving Forward by Looking Back

JESSE C. RIBOT

> ... Somos muitos Severinos
> iguais em tudo e na sina:
> a de abrandar estas pedras
> suando-se muito em cima,
> a de tentar despertar
> terra sempre mais extinta,
> a de querer arrancar
> algum roçado da cinza.
> Mas, para que me conheçam
> melhor Vossas Senhorias
> e melhor possam seguir
> a história de minha vida,
> passo a ser o Severino
> que em vossa presença emigra.
>
> João Cabral de Melo Neto, *Morte é Vida Severina*, 1966

Severina is a common name in the dry Northeast of Brazil – it is a name akin to the word for severity, reflecting rural life in these drylands. The experience of severe deprivation is widespread in the semi-arid tropics. Indeed, there are many Severinos living lives of constant vulnerability to hunger, famine, dislocation and material loss. Extreme climatic events, such as droughts, simply unveil this underlying chronic state. Such vulnerability is not caused by climate variability or climate change alone (Sen 1981; Watts 1983*a*). It is a result of the configuration of forces that shape the ability of farm and pastoral populations to produce, reproduce and develop. Climate extremes are an expected characteristic of semi-arid lands. Most populations know from local history the frequency and likely consequences of extreme climatic events. And most populations in highly variable or extreme climatic zones shape their livelihood systems to buffer against potential catastrophes. They prepare with the means at their disposal for these expected, yet unpredictable, threats. But why are some livelihood systems, some populations, some regions and some socioeconomic groups less able to prepare for or recover from natural extremes? What shapes their exposure to and ability to

rebound from these expected events? What shapes their vulnerability in the face of climate variability and change?

This volume explores the relation between climate variability or climate change and the wellbeing of rural agricultural and pastoral populations in the semi-arid tropics. We focus on climate-related vulnerabilities because climatic events in these lands, particularly droughts, trigger frequent subsistence crises – sharply increasing crop failures, dislocation, hunger and famine. The cycle of drought also, through the dislocation process and impoverishment of some and through speculative profits and concentration by others, increases stratification and the social inequities that help shape chronic deprivation and vulnerability (Watts 1983*a*). This volume focuses on the semi-arid tropics – those semi-arid lands in the developing world – because it is here that the crisis is most acute and where the need for change is most immediate. While there is vulnerability to material loss in the temperate semi-arid lands – such as the southwest United States, no one starves when drought strikes. But drought of the same magnitude in the semi-arid tropics of Northeast Brazil or the Sahel is likely to result in hunger or famine, and would certainly result in widespread human misery. Indeed,

1

it is in the semi-arid tropics where drought and poverty collide. It is here that the processes of underdevelopment undermine the coping abilities and resilience of entire populations. This introduction focuses on how the contributing authors understand the causes of vulnerability, because it is through understanding these causes that the impacts associated with climate can ultimately be reduced.

The chapters of this volume were drawn from the more than 70 papers presented at the International Conference on the Impacts of Climate Variability and Sustainable Development in Semi-Arid Regions (see Preface). The papers were chosen for their broad historical perspectives on vulnerability and social response as well as for a wide geographical representation. The conference was an effort on the part of social scientists, climatologists, policy analysts and policy makers to examine, find solutions for and bring attention to the common problems faced by the peoples of semi-arid lands – the most dramatic of which are associated with climatic phenomena. Conference participants were asked to examine the past consequences of and responses to climate variability and, given these past experiences, to reflect on the likely effects of and possible proactive policy responses to the types of climatic events that global warming might bring. Nations and peoples of the semi-arid regions of the world have long histories of planning for, coping with, rebuilding after, and responding to variations inherent in their climates. The chapters in this volume recount some of these histories and reflect on the future of the semi-arid lands of the least-developed countries.

The stories in this volume follow different Severinos through different places and times – the Tlaltizapan region of Mexico, Machakos district of Kenya, Northwest China, Australia's Eyre Peninsula, Northeast Brazil and Amazonia, and the Midwest United States – to better know the various aspects of vulnerability and responses to it. These case studies address our questions by examining past experience so that we can respond now to reduce the vulnerability of current and future generations.

THE CAUSAL STRUCTURE OF VULNERABILITY

Climate impact analysis is a way of looking at the range of consequences of a given climatic event or change (Rosenzweig and Parry 1994; Rosenberg and Crosson 1992). For instance, drought, however defined, is associated with a number of outcomes. Impact assessments begin by mapping out direct physical consequences of a climatic event or change, such as: reduced crop yields; livestock losses; reservoir depletion; hydroelectric interruptions; drinking-water shortages or quality changes; dryland degradation; or forest

dieback. These direct outcomes can then be traced to social consequences, such as: forced sales of household assets or land; further ecological losses such as soil erosion and deforestation from people trying to cope with the other losses by drawing more heavily on their remaining natural resources, speculative price rises driving food prices out of the range of the poorest households, dislocations resulting in out-migration to shanty towns and farther frontiers, destitution or servitude; disease outbreaks; hunger; or famine.

But it is misleading to designate these outcomes as 'impacts' of climate variability or change. This type of impact analysis implicitly attributes to nature causality that can be directly and more productively traced to social organization. 'Vulnerability' analysis (see Downing 1991, 1992; Ribot, Najam and Watson, this volume) provides a basis for tracing out social causality. Vulnerability analysis turns impact analysis on its head by examining the multiple causes of critical outcomes – dislocation, hunger, famine – rather than the multiple outcomes of a single event (Downing 1991, 1992). It traces outward from each instance of vulnerability the multiple physical, social and political-economic causal agents and processes. In doing so, it places climatic events among the social and political-economic relations and processes that shape the negative consequences with which we are concerned. By linking climate-associated 'impacts' (or outcomes) to causality, vulnerability analysis can also provide a sound basis for policy, since it is through responding to its causes that vulnerability can be redressed.

The methods for analysis of vulnerability – developed from the work of Amartya Sen (1981) – lay the groundwork for examining causality in a systematic way. Sen (1981, 1987), examining vulnerability to hunger and famine, begins at the household level with what he calls entitlements. A household's food entitlement consists of the food that the household can obtain through production, exchange, or extra-legal legitimate conventions such as reciprocal relations or kinship obligations (Drèze and Sen 1989). A household's assets, or endowment (cf. Drèze and Sen 1989), include investments in productive assets, stores of food or cash, and claims on other households, patrons, chiefs, government or on the international community that a household can make (Swift 1989:11). 'Assets create a buffer between production, exchange and consumption' (Swift 1989:11). They form the basis of a household's entitlements. In turn, assets depend on the ability of the household to produce a surplus that they can store, invest in productive capacity and markets, and in the maintenance of social relations (cf. Scott 1976; Berry 1993).

Vulnerability in this framework is the risk that the household's entitlements will fail to buffer against hunger, famine, dislocation or other losses. It is a relative measure of the household's proneness to crisis (Downing 1991).[1] The power of this analytic framework is that from each instance of entitlement failure – i.e. each instance in which a household's production, investments, stores and claims are insufficient to sustain them – chains of causality can be traced out. From an understanding of causality, we can move toward policies to reduce the vulnerability which is located at the confluence of these causal chains (cf. Jessop 1990:13).

Through a household perspective on vulnerability, climatic and other environmental phenomena can be understood socially. By focusing on the household, Sen (1981), concerned with entitlement failure, and Blaikie (1985), concerned with environmental change, have replaced ecocentric models of natural hazards and environmental change with social models in which environmental fluctuations and changes – in climate, forests or soils – are located among the other material and social conditions shaping and shaped by household wellbeing (see Watts 1983b). By incorporating environment (including climate) into a social framework, the environment may appear to become marginalized – set as one among many factors affecting and affected by production, reproduction and development. But this does not diminish the importance of environmental variability and change. Indeed, it strengthens environmental arguments by making it clear how important, in degree and manner, the quality of natural resources is to social wellbeing. These household-based social models also illustrate how important it is that assets match or can cope with or vary with (as in buffer against) these environmental variations and changes so that the activities of production and reproduction on the land are not undermined by and do not undermine the natural resources on which they depend.[2]

Watts and Bohle (1993), extending Drèze and Sen's (1989) analysis of entitlements, argue that vulnerability is configured by the mutually constituted triad of entitlements, empowerment and political economy. Here, empowerment is the ability to shape the political economy that in turn shapes entitlements. Drèze and Sen (1989:263) have empirically observed the role of certain types of political enfranchisement in reducing vulnerability, particularly the role media can play in creating a legitimation crisis in liberal democracies. Watts and Bohle have situated Drèze and Sen's analysis in a broader theoretical framework, allowing for a more systematic analysis of the political economy of the production and reproduction of vulnerability. Their analysis illuminates the role of empowerment through enfranchisement in redressing the inequities ongoing political-economic processes produce.

The chapters in this volume, by examining the causes of vulnerability in their historical context, place vulnerability in its broader social and political-economic context. While they do not use the terminology of vulnerability (e.g. entitlements, enfranchisement, empowerment, capabilities), they show

causal relations in the historical and contemporary production and reproduction of vulnerability at the level of the state, region, community and household. They show drought, floods, commodity price fluctuations and conflict as triggering crises, while locating the causes of vulnerability in the face of these events in the longer-term production of vulnerability through state policies, class relations, and demographic shifts. The chapters by O'Brien and Liverman, Wang'ati, Zhao, Glantz, Heathcote, and Bitoun, Guimarães and Araújo show how state policies – ranging from spatial and class inequalities in colonial and contemporary development, land-tenure and economic arrangements, to supports for the occupation of marginal lands – have played major roles in both security and vulnerability through their effects on resource access and population movements. They touch on the ways in which state development policies and processes have relegated some regions or classes to higher vulnerability while aiding others. In their cases, vulnerability has been shaped through the allocation of state funds and through the structuring of class-differentiated access to alternative income-generating opportunities and access to productive resources such as land, irrigation, credit, fertilizer and improved seed. O'Brien and Liverman, Glantz, and Heathcote also discuss the production of vulnerability through forces pushing and pulling people of particular classes down the rainfall gradient into more and more marginal lands, while Wang'ati cites population growth (in the absence of alternative employment opportunities – cf. Drèze and Sen 1989), in addition to population movements, as contributing to vulnerability. But, while state action and inaction, in addition to broader forces of differentiation and marginalization, may be responsible for producing much of the vulnerability faced by marginal populations, Ribot, Najam and Watson, Wang'ati, Zhao, Glantz, Heathcote, and Wilhite also argue that it is through state intervention that vulnerability can ultimately be reduced.

Taking crises that are often construed as impacts of climate extremes and tracing their causes, through vulnerability, to the social and political-economic system in which agricultural and pastoral households are located is a first step in moving toward durable policy responses. The chapters of this volume, outlined below, take a major step in the direction of linking climate-associated crises to social cause, and hence, potential response.

ORGANIZATION AND CONTENTS

The volume is divided into four parts. Part I is an overview, outlining the issues facing the populations of semi-arid lands. Part II consists of four country case studies examining the past responses to climate variability and potential consequences of climate change. Part III includes three chapters with diverse objectives but all exploring the historical development – that is the causes – of climate-related vulnerabilities, or strategies for reducing the vulnerability to and impacts of extreme climatic events. Part IV contains the Declaration of Fortaleza, hammered out by the participants of the International Conference on the Impacts of Climate Variability and Sustainable Development in Semi-Arid Regions (ICID), and the working group summaries that served as a starting point for conference-wide discussions. The volume and the papers within it are outlined briefly below.

Part I is a synthesis chapter, 'Climate variation, vulnerability and sustainable development in the semi-arid tropics,' in which Jesse C. Ribot, Adil Najam and Gabrielle Watson tease out some overarching themes from the 76 papers and presentations at ICID. The authors focus on the relation between impacts and vulnerability and on the policy ramifications of vulnerability analysis. Impact assessment can indicate the types, distribution and magnitude of consequences of climatic events. It does not, however, automatically lead to relevant policy recommendations unless it is coupled with an analysis of the social causes of climate-related vulnerability. The object of vulnerability analysis is to bridge the gap between impact analysis and policy formulation by directing policy attention to the root causes of vulnerability rather than to its symptom – the negative outcomes, 'impacts', that follow triggering events such as droughts. This chapter also reviews and discusses definitions of drought, dryland degradation, desertification, climate variability and climate change, as well as the main problems involved in projecting climate change and its implications. In addition, the chapter sketches the main technical and institutional policy options that appear in the papers from the conference.

Part II, 'Climate variation, climate change and society,' consists of case studies of past responses to climate variability and potential consequences of climate change in Mexico, the countries of the African Sahel, China and Australia. In this section's first chapter 'Climate change and variability in Mexico,' Karen O'Brien and Diana Liverman discuss the increase in vulnerability to the consequences of drought under colonial rule, and its persistence in present agricultural arrangements. They examine the probable consequences of global warming with attention to the social underpinnings of climate-related vulnerability.

O'Brien and Liverman provide evidence linking past famines with drought, social unrest, speculative price rises, colonial economic and land-tenure arrangements, and the colonial political economy. They then show that current vulnerability is a function of the different agricultural practices associated with farm size – a relationship rooted in the colonial land-tenure system – hence linking vulnerability with political economy. They point out that the worst effects of drought have been observed in the rainfed agricultural areas cultivated by the poorest *ejidatarios* and *campesinos*, who have limited access to credit, irrigation, fertilizers and improved seeds. Hence, past and present vulnerability is closely linked with the political-economic circumstances faced by different classes of farmers in different geoclimatic regions, and small farmers tend to be the most vulnerable group.

O'Brien and Liverman's research also shows that drought-related crop losses have increased over the past four decades. This increase is only partly due to the mere extension of cultivated area, since the percentage of losses has also increased with time. The increased percentage probably reflects the expansion of agriculture onto more marginal land, since these losses have not been associated with any increase in the intensity of weather extremes. Thus, these increased losses probably reflect increased vulnerability for a particular class of small farmers on marginal lands (cf. Glantz, this volume; Heathcote, this volume).

The authors also compare the five most prominent climate change projections for Mexico and systematically examine their limitations and implications. They apply the climate model results to a more detailed study of the Tlaltizapan region, and find that 'in spite of differences between the climate model results, the general direction of changes in Mexican maize rainfed and irrigated yields with global warming is a decrease, whatever the model. Sensitivity analysis indicates that decreases in crop yields will be severe under global warming unless irrigation expands, fertilizer use increases, or new varieties are developed.' The ramifications are clear. If we expect the climate to warm and if we have reasonable confidence in the climate models, the 'unless' must be taken seriously. But, even without climate change, as the authors demonstrate through historical analysis, implementing such measures to reduce vulnerability is already needed.

Fredrick J. Wang'ati's chapter, 'The impact of climate variation and sustainable development in the Sudano-Sahelian Region,' examines past climate-related trends and experiences, the evolution of socioeconomic indicators in the region, and proposes a strategy to increase ecological sustainability and reduce subsistence vulnerability through locally based organization and innovation. In the Sudano-Sahelian region, the dry and highly variable climate is a major factor shaping ecological and socioeconomic processes. Wang'ati locates the causes of climate-related vulnerability in population growth, reduced mobility of pastoral groups, international conflict and changes in land-tenure regimes. He also cites inadequate traditional technologies, a low level of economic development and the lack of alternative income-generating opportunities as contributing factors. These factors have conspired to push permanent settlements onto previously only intermittently used marginal land.

According to Wang'ati the trends in Kenya's 'key variables' show that rainfall is declining, forest resources, food production and economic indicators are on the decline, and import dependence is increasing. But in the midst of these ominous macro trends, Wang'ati examines the very successful case of development in Machakos district of Kenya. Local-level analysis reveals that productivity and ecological management have evolved favorably over the past 60 years. In Machakos in the 1930s colonial officials doubted whether the district could feed itself. Since then the growing population has evolved a complex farming system which appears to be sustainable and even produces enough to export some products. Wang'ati attributes these positive outcomes to the slow and systematic evolution and development of locally based farming and pastoral systems along with the emergence of farmers' organizations. The case demonstrates the capacity for some semi-arid lands to support greater populations, increase productivity and contribute to the national economy. It also could support the hypothesis that population pressure can lead to innovation and development, rather than ecological collapse (Boserup 1988). It is interesting to note how macro-level variables were unable to capture the local-level diversity and dynamics that might contain the seeds for success on a broader scale. The inconsistency in this case illustrates how important it is to choose the scale of analysis, as well as intervention, with great care. Relying on overly general indicators alone can mask both problems and potentials.

From this and other cases, Wang'ati draws some recommendations for what he calls a 'sustainable development strategy' (SDS). His strategy follows from the simple tenets that sustainable development must build on existing successful resource management strategies, and that development should proceed by progressive improvements on proven coping strategies. He argues that wholesale changes in land use and large-scale projects should be avoided in favor of gradual, or 'stepwise' community-based improvements. While Wang'ati does not use the language of vulnerability – entitlements, political economy and empowerment – these stepwise, locally based, locally organized improvements reflect the importance and role of empowerment. The people of Machakos are enfranchised to proceed as they see fit. They have been able to use this freedom to shape the institutions and activities that underpin their wellbeing.

On a national scale, Wang'ati asserts that SDS should include the creation of urban income-generating opportunities so that rural–urban migration can be a positive way to help relieve excessive pressures on semi-arid lands. In addi-

tion, development of alternative agricultural and non-agricultural rural livelihood systems, research on drought-resistant crops, improvements in marketing and transport infrastructure, increased interstate trade, greater rural retention of added value, and increased diversity of crop and economic activities are all part of Wang'ati's sustainable development recommendations. Again, these are all moves to support from the state's side the enfranchisement of rural populations – by providing access to resources and livelihood options. Through these opportunities, and with increased savings, local populations can build their entitlements, ultimately reducing their vulnerability.

In 'Climate change and sustainable development in China's semi-arid regions' Zong-ci Zhao characterizes the climatic conditions, natural resources, economic productivity and populations of China's semi-arid northwest. Zhao also outlines some major changes in these variables and climate-related problems facing development of the region. This chapter is unusual in that Zhao has at her disposal temperature measures dating back 500 years and information on drought and floods for more than 2000 years. Her recent data show that in the last 40 years the local climate has warmed between 0.5 and 1.8°C. Regional climate modelling exercises show that warming will continue; however, as in other chapters, the uncertainties are too great to draw conclusions.

Zhao demonstrates that China's semi-arid lands are harsh. She leaves us with the somewhat pessimistic regional adage: 'Eat your fill in summer, grow fat in autumn, get thin in winter, and die in spring.' But, she also leaves us with the sense that the region has enormous potential for development. China's semi-arid regions have been a target for development as the central government regards them as a population pressure valve for the overpopulated South and Southeast (quite the contrary of the role of Brazil's semi-arid Northeast: see Bitoun, Guimarães and Araújo, this volume). As such, government policies have encouraged migration into this region. Zhao points out that not only are there solar, wind, hydro and mineral resources, but there are enormous expanses of uncultivated land into which she suggests farming could be expanded. There are also great expanses of grazing land. But to use them sustainably could be risky without appropriate measures. Indeed, deterioration of currently utilized grasslands of Northwest China still must be turned around through water and soil development and conservation.

This chapter compels us to ask how vulnerability to crop failures, hunger and famine are different in a centrally planned nation where land is presumably already equitably distributed (cf. Przeworski 1990). How does the social structure of vulnerability differ from non-centrally planned Mexico, the African Sahel and Brazil's Northeast? Whose decisions shape the relative vulnerability of different social groups and livelihoods within semi-arid China, and how? Zhao's comments also indicate how critical government decisions are in making these 'marginal' lands more productive in marked contrast to the examples given by Heathcote and by Glantz in their chapters. Here is a clear example of the critical nature of state political economy, through policy decisions, in shaping vulnerability or security in these lands. In the language of vulnerability, a centralized state control over political-economic relations of access and development would be inversely related to the degree of local empowerment and enfranchisement in shaping local entitlements.

R. Les Heathcote, in his chapter 'Settlement advance and retreat: a century of experience on the Eyre Peninsula of South Australia,' tells the story of the clearing and occupation of South Australia's semi-arid Eyre Peninsula, and of responses to and perceptions of climate-related land use and land quality changes. By carefully reviewing the past 100 years of public records and evaluating data available on climatic, ecological and economic conditions, he brings out the primary variables involved in sustaining the region's agricultural viability. In this region, periodic droughts associated with the fluctuating economic fortunes of a region producing grains and livestock for an international market have created stresses which have led, in part, to conditions that exacerbate the desertification process. Governments have played various roles, ranging from that of real-estate broker through to steward of the national environment. The general trend has been toward increased and more centralized government control over land use. These state policies have generally supported the family farm, which remains the region's basic form of productive unit. Heathcote argues, however, that growing costs of production and diminishing real income have pushed producers to gamble increasingly with the seasons and with the ability of the environment to recover from drought-related deterioration.

Heathcote's study shows that farm households are vulnerable to ecological decline – that is the undermining of their productive entitlements, of which the ecological base is one – in the face of economic as well as climatic fluctuations. Farmers on the Eyre Peninsula had to increase the area farmed or intensify farming during periods of low farm-product prices. Here, farm viability is a function of the crop yield per hectare necessary to break even. When prices drop so low that yields must exceed the sustainable productive capacity of the land, debts are incurred, forcing farmers to intensify farming and 'mine' the land.

This chapter locates the causes of farmers' vulnerability and related land degradation – which occur in the face of fluctuations of factor markets and climate – somewhere between state policies supporting settlement on these lands and the risk-taking these policies facilitate. Contrary to the notion that state support and development initiatives reduce vulnerability, Heathcote's case study implies that state support encouraged farmers to settle these marginal lands – supporting Glantz's arguments (this volume). The assumption of part of the risk by government reduced risk to the farmer while increasing the potential economic losses to society, or the nation as a whole. While some types of vulnerability may have increased, it is possible that vulnerability to hunger, famine and dislocation were reduced. Whether the economic losses associated with farm failures in this region were worth the broader economic and social benefits of producing on these lands appears still to be an open question. It is also an open question as to why government intervened here in the first place. Is this an instance of a relatively empowered and enfranchised population being able to make claims on government, shifting risk from themselves and their bodies – as in risk of hunger or famine – to a more economic form of risk shared by government institutions?

Part III of this volume, 'Climate variability and vulnerability: causality and response,' includes chapters exploring the causes of climate-related vulnerabilities, as well as past and potential policy responses. The first of these chapters, 'Drought follows the plow: cultivating marginal areas,' by Michael Glantz, argues that contrary to beliefs of the past that rain would come to those areas brought under cultivation, new lands brought under the plow are drought-prone. Glantz argues that this is because all the world's best rainfed agricultural land is already in use. Hence, the only new lands available for farming are, by necessity, lower quality and higher risk. Putting forth this hypothesis, Glantz then describes conditions, illustrated by cases in Africa, Brazil and Australia, under which populations end up in the precarious position of farming drought-prone lands. The conditions Glantz describes include 'push' and 'pull' factors. Population growth, environmental degradation and particular government policies constitute pushes into more marginal, dryer lands, while the attraction of good growing conditions during unusually wet periods constitutes the pull.

As do O'Brien and Liverman and Heathcote in their chapters, Glantz is challenging the relative importance of changing climate as a causal factor in *increased* incidence of drought and its associated consequences. He is exploring the relative importance of social effects in shaping the negative outcomes usually associated with climate. Glantz is effectively posing the following hypothesis: What is being read as increased frequency and intensity of droughts is really an indicator of greater numbers of people farming more marginal lands – that is, the location of human populations with respect to climate resources helps explain the increased frequency and magnitude of negative climate-related outcomes better than do characteristics of climate variability or change. This formulation directs us to examine more carefully why populations move into these lands, rather than why these lands are dry. In short, he is proposing a chain of causality of increased vulnerability to climate-related disasters that begins with social forces pushing or pulling populations into marginal lands and ends with exposure to the pre-existing conditions of those lands. Hence, to understand the increased vulnerability of these populations we must understand the roots of their need to eke out a living at this particular margin.

Glantz's formulation begs an additional question, partly addressed by Heathcote's exploration here of the effects of state policies in Australia and also provoked by Zhao's assertion in her chapter that agriculture *should* be expanded onto currently uncultivated lands. The questions are: What conditions are necessary to farm and raise livestock in these dryer, and hence more drought-prone, regions with less risk? What kinds of investments are necessary? What kinds of social security are needed? These questions must be asked, since movement into these areas is under way, and this is one place where productive action can be taken. Does Zhao's recommendation have to increase vulnerability, as in Heathcote's and Glantz's cases, or can measures be taken that assure low levels of vulnerability in these dryer lands. With proper state interventions, or local initiatives, as shown by Wang'ati's case study, secure and sustainable farming could perhaps be integrated into these regions. This latter question links the causes of vulnerability to the *lack* of action – state and local – and the social and political-economic conditions that systematically bias against or prevent investment, and producer organization and innovation, in marginal lands occupied by marginalized populations. That is, what shapes state or private investment in these lands and what prevents or facilitates those living there using their resources, knowledge and capabilities to reduce their own vulnerability through adaptation and innovation (see Deere and de Janvry 1984; Blaikie 1985)?

In essence, Glantz's formulation points out that marginality and drought-proneness are partly inherent characteristics of *place*, with the social forces and causes of vulnerability being linked to the reasons people arrive in these places. Yet the types of social organization – such as the cooperative farming in Machakos described by Wang'ati – or the availability of infrastructure, ranging from irrigation and extension services to marketing facilities, social security systems or credit, are also essential determinants of whether these 'marginal' lands can be productively and securely farmed (cf. Zhao, this volume; Wilhite, this volume). Perhaps, once again, this is a function of how well the populations of these lands can retain their income locally, or make claims on state institutions – both of which are issues of economic and political enfranchisement.

Jan Bitoun, Leonardo Guimarães Neto and Tania Bacelar de Araújo's chapter, 'Amazonia and the Northeast: the Brazilian tropics and sustainable development,' examines the history of the relation between Brazil's semi-arid Northeast and humid Amazonia – two regions linked more by their geographical, economic and social marginality within Brazil than by their physical characteristics. The authors argue that Brazil's more economically developed Southern and southeastern region shaped development in the Northeast and Amazonia, as well as the relation between them, through systematic biases in the perception of the national role of, and hence the policies toward, these regions. The authors locate the causes of environmentally unsustainable relations between peasants and the land in the inequitable development policies resulting from these political, economic and ideological biases.

The authors make four critical points in their analysis. First, sustainability of the Amazon cannot be considered separately from the broader geographical, social and political-economic context in which the Amazon is located; part of this context is Amazonia's close relation to the semi-arid Northeast from which much of Amazonia's population has migrated. Second, the Amazon and Northeast Brazil are integrally linked, since without proper development in the Northeast – capable of curbing out-migration – the flow of colonizers from the Northeast into the Amazon will continue to undermine the Amazonian economy and ecology either directly or via the South and Southeast. Third, unequal national-level development policies have supported inequitable relations of production in these two regions and have undermined, rather than supported, ecologically sustainable practices. Last, the conjuncture of recent developments in awareness of the importance of locally rooted development, national and international attention to ecological concerns in the development process, and recent media focus on Amazonia, has created an opening in which to examine development strategies suitable to the Amazon and Northeast and compatible with the national development agenda. But the authors warn that development in the region will not be sustainable until it addresses land distribution and labor issues, and challenges the deeply rooted agrarian structures that remove surpluses from local producers, thus precluding the reinvestment of resources into the maintenance and development of ecologically sound agricultural practices. In addition, they point out that ecologically sound development proposals 'will not be viable if political and interregional considerations are not analyzed, along with the economic and technological criteria.'

These authors locate the causes of and responsibility for environmental decline and human vulnerability at the level of national political economy. They also show that the lack of development in one region – which they attribute to geographically inequitable national development policies – can have profound social and environmental repercussions on other regions. This implies that social vulnerability and environmental decline must be addressed on the geographical and political-economic scales that shapes them. It also implies that the vulnerability of regions is interconnected: those interested in preserving the Amazon will have to attend to the development of surrounding regions if they are to stem the influx of landless farmers.

Donald A. Wilhite's chapter, 'Reducing the impacts of drought: progress toward risk management' documents an important transition taking place in government response to drought from one of crisis management to one of risk reduction. After illustrating this transition with cases from the United States of America, South Africa and Australia, he recommends a procedure for developing and implementing integrated drought policies and plans. The chapter demonstrates the importance of proactive planning for reducing drought impacts, and the relatively low cost of planning compared with the high economic, human and environmental costs of a late or inadequate response. Wilhite's proactive program develops an approach to crisis planning and touches on the notion of vulnerability reduction. This is an appropriate chapter to close this section, since it is about where to go from here. It is about points of entry into reducing both vulnerability to and impacts of the expected but unpredictable normal variations of our climatic resource.

Part IV of this volume, 'The International Conference on the Impacts of Climatic Variations and Sustainable Development in Semi-Arid Regions,' presents the collective efforts of over 600 social scientists, physical scientists, politicians and diplomats from 46 countries to state clearly the common problems, needs and prospects of the semi-arid regions of the world. These efforts are embodied in the 'Declaration of Fortaleza' and the findings of the ICID Working Groups into which conference participants divided and from which the first draft of the Declaration was formulated. The Declaration was circulated and a final draft hammered out in an animated set of debates in the final plenary meeting. The contents of these documents are too varied to flesh out here, but they speak for themselves, and the participants of ICID hope they also speak to the concerns of the peoples who live their lives in semi-arid regions around the world.

USES OF HISTORY: PUTTING VULNERABILITY IN THE PAST

Vulnerability to hunger, famine, dislocation or material loss results from the dynamics of the social system in which agricultural and pastoral households are located. Vulnerability is shaped by ongoing processes of social differentiation and marginalization, within a specific social history of access to productive resources, formal and informal social security arrangements, state development policies, conflicts, etc. The resulting distribution of material stocks and of access to income opportunities, land and other material resources, as well as access to formal and informal social security arrangements, spells out the material and social conditions circumscribing vulnerability for some households and security for others. Climate extremes may trigger subsistence crises, but these crises come at the confluence of historical processes, as well as actions and events that make individuals, households, enterprises or regions vulnerable.

Vulnerability must be examined with an eye to vulnerability reduction. First and foremost, vulnerability must be understood by examining its historical antecedents – this is an essential part of the switch from impact analysis to vulnerability analysis (Downing 1991; also see Ribot, Najam and Watson, this volume). In examining historical antecedents we must ask: What types of processes and events cause vulnerability? What can be done to reverse deleterious processes and buffer against crisis-triggering events? Which processes are reversible, and which are not? We must also ask: Which processes can, both technically and politically, be modified and which cannot? Historical analysis reveals the underlying causes of vulnerability, but not all these causes are tractable. Here we need to reflect on the uses of history, rather than just retreating into proximate analyses that ignore history, or falling into paralysis by insisting on redressing all of the deepest causal relations at once. For example, early warning of the onset of hunger and famine does not have to focus on the moments before bodies begin to drop. Relief does not have to come in to avert only the most acute and horrifying crises. Identifying vulnerability gives us warnings of acute crisis even years in advance. Understanding vulnerability should thus be used to produce more durable and earlier, proactive responses. The uses of history in developing these understandings and in formulating responses needs further exploration.

Enfranchisement and empowerment must be part of any pathway for ushering vulnerability into the past. Security is about having the material resources – entitlements – to buffer against contingencies. However, as Drèze and Sen (1989) argue, it is more about the *capability*[3] to achieve food security and other valued 'functionings,' such as attaining sufficient entitlements. Capabilities are partly predicated on the rights that allow the pursuit of such wellbeing. The concept of capabilities, as Watts and Bohle (1993) state, 'approaches the totality of rights by which individuals, groups and classes command endowments and commodity bundles.' It is this formulation of capabilities that links the material with the political-economic aspects of vulnerability and security. Capabilities, through their basis in freedoms and rights, link entitlements with enfranchisement. In this manner, enfranchisement becomes a means by which material wellbeing can be pursued and maintained through intervention in political-economic relations and processes.

The entitlement approach complemented by empowerment and political economy reveals two complementary directions for vulnerability reduction: (1) redistribution (of ownership and access) and development in the material realm, relating to the material basis for entitlements; and (2) enfranchisement in the political-economic domain, relating to the ability to shape political economy. The political-economic relations that produce vulnerability by structuring access to resources and affecting maldistribution through processes of social differentiation and marginalization, could be directly confronted by legal reforms over access and redistribution of material wealth. But, because the processes of differentiation, stratification and marginalization are ongoing, there needs to be a countervailing set of processes, not just one-time moves to shift access control or redistribute assets. The process of overlapping spatial and social marginalization – marginalizing particular classes of people onto marginal lands – is, for example, a theme that runs through the chapters of this volume. In response, there needs to be a constant process in which those being marginalized are themselves constantly able – capable, that is – to reincorporate themselves and make claims that redress the constant production of their disadvantaged position. Enfranchisement and empowerment provide such ongoing countervailing forces. Enfranchisement and empowerment, along with other countervailing mechanisms and forces, are areas still wide open for theoretical development and empirical research – if vulnerability, rather than the people of the semi-arid tropics, is to be denied a future.

While this introduction focuses on the causes of vulnerability as they impinge upon the household, other units of social aggregation, such as the community, enterprises, civil organizations, markets, the state, or regions, can also be vulnerable, and can also be subject to a similar analysis of the causes of their vulnerability. Vulnerability exists at all levels of society. In addition, vulnerability of one segment of society influences vulnerability of others. Vulnerable households can render the state vulnerable to a legitimacy crisis, or to a costly set of state emergency interventions. Indeed, a state vulnerable to legitimacy crisis may provide a receptive forum for households to press the state for vulnerability-reducing goods and services. Each institution must be understood in the set of relations in which it is embedded; this includes its relation to the vulnerability and security of other groups within society. This relation among institutions or regions (see Bitoun, Guimarães and Araújo, this volume), and the interdependence of their security, is another important area for further study.

Vulnerability analysis is a good starting place for rethinking rural development.[4] It is local by definition: it builds outward from the household or whatever specific institution it focuses on (cf. Vayda 1983; Blaikie 1985; Watts 1987; Schmink 1992). Yet, while both the building and failure of entitlements occur concretely at this local level, a focus on vulnerability systematically includes the multiple temporal, spatial and social scales that impinge on the production and reproduction of everyday life. In this manner the local is not conceptually isolated from the global, and the analysis of vulnerability at each level is constantly referenced back to the security and wellbeing of those at risk. It is this multilayered analysis that traces causality outward, through space, through social relations and through history, from which an understanding of the conditions for secure and productive development can grow.

The semi-arid tropics are filled with Severinos whose lives are harsh even when there are no extreme droughts or floods and no changes in natural conditions. It is this chronic condition of deprivation that must ultimately be addressed. Climate variations and extremes are not new – and they may not even be changing. But, as Glantz argues in this volume, disasters associated with them will be more frequent and more intense as more people occupy more marginal lands. O'Brien and Liverman also point out in their chapter that these disasters will increase as the conditions of marginalized populations already occupying semi-arid lands for generations deteriorate – unless development of productive capacities proceeds in a timely manner. Those interested in analyzing or mitigating the 'impacts' of climate in semi-arid lands must begin through attention to causes of and responses to deprivation, underdevelopment, poverty and other aspects of vulnerability. Underdevelopment, poverty and vulnerability are also not new. There has been much scholarly thinking on these subjects. It is time to take advantage of what is already well understood and apply it to the crises of the present.

Ironically, this volume was launched by concerns articulated by those focusing on *future* climate change. Projecting climate change into the future, to a time distant enough to be free of commitment to immediate action or change, has brought to world attention a chronic tragedy taking place in the semi-arid tropics today. But hunger, famine, dislocation and material loss can still best be understood and redressed by learning from the past. By enabling current populations to buffer themselves against today's climatic variations, they will be better able to cope with future contingencies. Rather than focusing on the distant future, we must use the opportunity that the specter of climate change brings to slide back down the projection lines to point to and address the crisis at hand.

ACKNOWLEDGMENTS

I would like to express my gratitude to Alayne Adams, Jean Drèze, Michael Glantz and Michael Watts for their comments on drafts of this Introduction.

ENDNOTES

1. Downing (1991, 1992) develops the concept of vulnerability (for a summary see Ribot *et al.*, this volume). Watts and Bohle (1993:46), who take the concept further, define vulnerability as follows: 'Vulnerability is a multi-layered and multi-dimensional social space which centers on the determinate political, economic and institutional capabilities of people in specific places at specific times. In this sense a theory of vulnerability should be capable of mapping the historically and socially specific realms of choice and constraint – the degrees of freedom as it were – which determine exposure, capacity and potentiality. In a narrow sense this is about individual command over basic necessities; in a still broader sense it also speaks to the structural properties of the political economy itself.' Also see Chambers (1989:1), who defines vulnerability as follows: 'Vulnerability refers to exposure to contingencies and stress, and difficulty in coping with them. Vulnerability has two sides: an external side of risks, shocks, and stress to which an individual or household is subject; and an internal side which is defenselessness, meaning a lack of means to cope without damaging loss.'
2. Household models are often limited by their failure to account for intra-household dynamics of production and reproduction, but they do not have to be (see, for example, Guyer 1981; Guyer and Peters 1987; Carney 1988; Hart 1992; Agarwal 1993; Schroeder 1992).
3. Capabilities are defined by Drèze and Sen (1989:12ff) as 'a set of functioning bundles, representing the various alternative "beings and doings" that a person can achieve with his or her economic, social and personal characteristics.' 'While the *entitlement* of a person is a set of alternative *commodity* bundles, the *capability* of a person is a set of alternative *functioning* bundles' (13ff). 'The focus here is on human life as it can be led, rather than on commodities as such, which are means to human life, and are contingently related to need fulfillment rather than being valued for themselves' (13). They go on to define capabilities as 'the extent of the freedom that people have in pursuing valuable activities or functionings' (42).
4. Also compare with farm systems theory as used by Blaikie (1985).

REFERENCES

Agarwal, B. 1993. Social security and the family: coping with seasonality and calamity in rural India. *Agriculture and Human Values* [Winter–Spring] 156–65.

Bernstein, H. 1979. African peasantries: a theoretical framework. *Journal of Peasant Studies* 6(4): 420–43.

Berry, S. 1993. *No Condition is Permanent: The Social Dynamics of Agrarian Change in Sub-Saharan Africa*. Madison: The University of Wisconsin Press.

Blaikie, P. 1985. *The Political Economy of Soil Erosion in Developing Countries*. London: Longman.

Boserup, E. 1988. Environment, population, and technology in primitive societies. In *The Ends of the Earth: Perspectives on Modern Environmental History*, ed. D. Worster, pp. 23–38. Cambridge: Cambridge University Press.

Carney, J. 1988. Struggles over land and crops in an irrigated rice scheme. In *Agriculture, Women and Land: The African Experience*, ed. J. Davidson, pp. 59–78. Boulder: Westview Press.

Chambers, R. 1989. Editorial/Introduction. Vulnerability, coping and policy. *IDS Bulletin* 20(2). Sussex: Institute for Development Studies.

Deere, C. D. and de Janvry, A. 1984. A conceptual framework for the empirical analysis of peasants. Giannini Foundation Paper No. 543, pp. 601–11.

Downing, T. E. 1991. Assessing socioeconomic vulnerability to famine: frameworks, concepts, and applications. Final Report to the US Agency for International Development, Famine Early Warning System Project, 30 January 1991.

Downing, T. E. 1992. Vulnerability and global environmental change in the semi-arid tropics: modelling regional and household agricultural impacts and responses. Presented at ICID, Fortaleza-Ceará, Brazil, 27 January to 1 February 1992.

Drèze, J. and Sen, A. 1989. *Hunger and Public Action*. Oxford: Clarendon Press.

Guyer, J. 1981. Household and community in African studies. *African Studies Review* 24(2,3): 87–137.

Guyer, J. and Peters, P. 1987. Introduction. Special issue on households. *Development and Change* 18: 197–214.

Hart, G. 1992. Household production reconsidered: gender, labor conflict, and technological change in Malaysia's Muda region. *World Development* 20(6): 809–23.

Jessop, B. 1990. *State Theory: Putting the Capitalist States in Their Place*. University Park: The Pennsylvania State University Press.

João Cabral de Melo Neto. 1966. 'Morte é vida Severina,' in *Morte é Vida Severina, é otros poemas em voz alta*, 32nd edn. Rio de Janeiro: José Olympio Press.

Przeworski, A. 1990. Could we feed everyone? *Politics and Society* 19(1): 1–38.

Rosenberg, N. J. and Crosson, P. R. 1992. Understanding regional scale impacts of climate change and climate variability: application to a region in North America with climate ranging from semi-arid to humid. Presented at ICID, Fortaleza-Ceará, Brazil, 27 January to 1 February 1992.

Rosenzweig, S. and Parry, M. 1994. Potential impacts of climate change on world food supply. *Nature* 367: 133–8.

Schmink, M. 1992. The socioeconomic matrix of deforestation. Paper presented at the Workshop on Population and Environment, sponsored by Development Alternatives with Women for a New Era, International Social Science Council, and Social Science Research Council, Hacienda Cocoyoc, Morelos, Mexico, 28 January to 1 February.

Schroeder, R. A. 1992. Shady practice: gendered tenure in the Gambia's garden/orchards. Paper prepared for the 88th annual meeting of the Association of American Geographers, San Diego, CA, 18–20 April.

Scott, J. 1976. *The Moral Economy of the Peasant*. New Haven: Yale University Press.

Sen, A. 1981. *Poverty and Famines: An Essay on Entitlement and Deprivation*. Oxford: Oxford University Press.

Sen, A. 1987. Research for action: hunger and entitlements. World Institute for Development Economics Research, United Nations University.

Swift, J. 1989. Why are rural people vulnerable to famine? *IDS Bulletin* 20(2): 8–15.

Vayda, A. P. 1983. Progressive contextualization: methods for research in human ecology. *Human Ecology* 11(3): 265–81.

Watts, M. J. 1983*a*. *Silent Violence: Food, Famine and Peasantry in Northern Nigeria*, pp. 231–62. Berkeley: University of California Press.

Watts, M. J. 1983*b*. On the poverty of theory: natural hazards research in context. In *Interpretations of Calamity*, ed. K. Hewitt, pp. 171–212. London: George Allen & Unwin.

Watts, M. J. 1987. Drought, environment and food security: some reflections on peasants, pastoralists and commoditization in dryland West Africa. In *Drought and Hunger in Africa*, ed. M. H. Glantz. Cambridge: Cambridge University Press.

Watts, M. J. and Bohle, H. 1993. The space of vulnerability: the causal structure of hunger and famine. *Progress in Human Geography* 17(1): 43–68.

Wisner, B. 1976. *Man-Made Famine in Eastern Kenya: The Interrelationship of Environment and Development*. Discussion paper No. 96. Brighton: Institute of Development Studies, University of Sussex.

I

Overview

1: Climate Variation, Vulnerability and Sustainable Development in the Semi-arid Tropics

JESSE C. RIBOT, ADIL NAJAM and GABRIELLE WATSON

INTRODUCTION

This chapter aims to capture the central issues that emerged from the papers, presentations and discussions at the International Conference on the Impacts of Climatic Variations and Sustainable Development in Semi-arid Regions (ICID), held in Fortaleza-Ceará, Brazil from 27 January through 1 February 1992 (see Preface). But, given the breadth and depth of the 76 papers and the wide-ranging discussions during the conference, this chapter could cover only a small subset of the issues that arose. We chose to focus on the plight of socially, politically, economically and spatially marginal populations in semi-arid lands, and the urgent need for environmentally sound and equitable development efforts. These themes recurred throughout the papers, presentations and discussions at the conference.

This chapter draws from the materials and information presented at the conference, as well as the broader literature where relevant. While the themes within this chapter are derived largely from the conference, the arguments are shaped – as could not have been otherwise – by the experiences and perspectives of the authors. We did not try to represent the scope nor the depth of the issues covered at the conference, but rather, to characterize the problems and opportunities, and to explore what we felt were the most pressing concerns within the semi-arid regions of the world.

Climate variability, natural resources and development in semi-arid regions

Vulnerability to dislocation, hunger and famine are the most critical problems facing the inhabitants of semi-arid lands. These regions are subject to extreme variations in their relatively scant seasonal and inter-annual precipitation, resulting in recurrent droughts and floods. Natural resources of semi-arid zones, such as timely water supplies, fertile soils,

vegetation and wildlife, tend to be scarce, and the existing resources are easily damaged by changes in precipitation patterns and by human action. Many of the semi-arid regions of the world, particularly the semi-arid tropics, are also characterized by subsistence vulnerability and insecurity for the large majority of their rural populations in the face of land degradation and climate variation. Vulnerability, social and geographic marginality, environmental change and dryland degradation are central, interlinked and chronic problems.

Semi-arid regions cover between 13% and 16% of the earth's land area, and are home to approximately 10% of the global population (Heathcote, this volume).[1] They exist in tropical, sub-tropical and temperate zones and fall within or encompass both developed and underdeveloped nations (see Fig. 1). In developed areas, the southwestern United States, and parts of the Western Plains of Canada and the periphery of the Australian desert are semi-arid (see, for example, Cohen et al. 1992; Rosenberg and Crosson 1992; Schmandt and Ward 1992; Heathcote, this volume). The semi-arid tropics encompass large portions of the least-developed regions on earth. Of the 22 countries of Africa's Sudano-Sahelian region 18 are among the world's least-developed nations (Wang'ati, this volume). Brazil's semi-arid Northeast is its most economically deprived region (Magalhães and Glantz 1992). Semi-arid tropics also include Mexico's Central Plateau, parts of Argentina, Chile and Uruguay, Central and South India, western China (22% of the country), and northern Mongolia (Sen 1987; Drèze and Sen 1989; Tie Sheng et al. 1992; Zhao, this volume). Many of these regions are highly prone to anthropogenic and climatically related environmental deterioration, while their populations are prone to hunger and famine. Indeed, it is at this conjunction of climatic variability and underdevelopment that human vulnerability and calamitous social dislocations are most likely to occur.

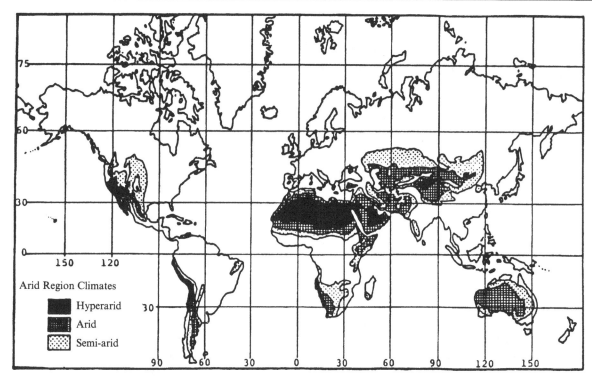

Figure 1: Arid and semi-arid regions of the world. (Source: Campos-Lopez and Anderson 1983:54.)

While rainfall, droughts and floods are physical phenomena, associated socioeconomic consequences (economic
failure, food shortages and outmigration) are linked to the
ability of affected populations to anticipate, prepare for and
respond to these events. The most striking characteristic of
the vast majority of the populations inhabiting the semi-arid
tropics is their lack of adequate human and financial
resources to cope with expected – even at times (with early-
warning systems) predictable – variability in their climatic
regimes (on early-warning systems, see Nobre *et al.* 1992;
Servain *et al.* 1992; Wang'ati, this volume). Because of poor
human, natural-resource and infrastructural development in
these regions, large portions of the population of semi-arid
tropics are vulnerable to hunger, famine, dislocation and the
loss of both property and livelihood in the face of climatic,
social, political or economic shocks. For the most part, their
lives are shaped by chronic job and food insecurity, inadequate, and in many cases non-existent, health care, low
wages, unemployment, under-employment and illiteracy, all
of which tend to amplify the social consequences of natural
phenomena.

Marginality and a low level of economic development
both exacerbate and are exacerbated by environmental
changes such as dryland degradation and deforestation.[2]
Exploited or marginalized populations are often excluded
from or bypassed by benefits of the development process,
and are pushed against their resource base, further eroding
its productive capacity. Mining the land (e.g. using land
resources in a manner that reduces productivity in the long

run) often becomes a necessity for those whose immediate
survival depends on these lands (Bernstein 1979; de Janvry
1981; Blaikie 1985; Blaikie and Brookfield 1987). Those who
are marginalized by the economic process onto the most
economically and ecologically marginal lands are the most
vulnerable populations. Under some conditions their marginality intensifies as they, and the marginal lands on which
they subsist, are exploited beyond their productive capacity.
The important question with respect to vulnerability in the
face of climate variability and change is why and how these
populations are marginalized and, hence, vulnerable. It is
this question that guides our attention to the social determinants of vulnerability.

Antonio Rocha Magalhães (1991:1) brings into focus both
who is vulnerable and why, when he characterizes the critical
nature of society's relation to the natural environment in
Northeast Brazil:

> Over the course of history, the economic, social and
> environmental impacts of adverse climatic events,
> especially droughts, have been calamitous. It is, however,
> the social dimension that accentuates the climatic
> problem in Northeast Brazil, as in many other developing
> regions. Here it represents a menace for the survival of a
> major part of the population because, unlike developed
> regions, the social agents are not equipped to face the
> consequences of adverse climatic events.

This chapter examines problems of the less-developed
semi-arid regions, because these regions are in most urgent

need of attention. Likewise, we focus on development as the path toward an environmentally secure and productive future.

Climate change

The regional consequences of anthropogenically enhanced global warming cannot yet be predicted with confidence. But some impacts are probable. Increases in temperature will result in an increase in evapotranspiration. This increase will be particularly significant in places where the climate is hot under current conditions. Whether rainfall in these regions will increase or decrease remains highly uncertain. But, the Inter-governmental Panel on Climate Change (IPCC 1990:iii, 12–13, and 20 ff.) indicates that semi-arid regions are among those areas most likely to experience increased climatic stress. Further, climatic change will have as yet unpredictable and perhaps unexpectedly extreme consequences with respect to frequency and intensity of precipitation and temperature variability for semi-arid regions.

Several regional climate-change scenarios designed to identify possible implications of global warming for semi-arid lands were generated by climatologists and social scientists and presented at the ICID conference. The tremendous uncertainty involved in projecting regional climate change is compounded by uncertainties in future productive capacities, demographic changes and socioeconomic development in these regions (see under Climate Variability and Change, below, for a discussion of these factors). There are nonetheless lessons that can be derived from climate change simulations and scenarios. For Mexico, O'Brien and Liverman (this volume) found decreasing soil moisture predicted by all climate-change models applied to Mexico. If soil moisture decreases in semi-arid regions, as can be derived from some general circulation models (GCMs) and is assumed in most of these scenarios, then productivity in these regions will most certainly decrease in the absence of considerable development efforts (Magalhães 1991; Downing 1992; El-Shahawy 1992; Schmandt and Ward 1992; Santibáñez 1992; Selvarajan and Sinha 1992; Cohen *et al.* 1992; O'Brien and Liverman, this volume). Most of the scenarios project worsening climatic conditions, in the form of more frequent droughts and shorter growing seasons. Some point to the possibility of a higher degree of inter-annual climate variability and of unexpected extreme meteorological events such as cool periods or more frequent floods (Izrael 1992:2–5). For these regions in which planning productive activities is already difficult due to the high climatic variability, climate change introduces even greater uncertainties and, thus, greater risks. A better regional understanding of climate change will help in planning for its consequences. But this is an insufficient policy response to the needs of semi-arid regions. To help cope with future regional uncertainties

generated by climate warming, policy makers must also address issues associated with current climate variability. Policies to address problems of populations living under current variability will be an invaluable basis for coping and adapting should climate change increase variability or drought.

Even assuming continued current climatic conditions, semi-arid regions may well be worse off in 10, 20 or 30 years due to the declining productivity of the land and increasing populations without access to alternative income-generating options (see, for example, Wang'ati, this volume). Indeed, the magnitude of natural hazard losses has increased in the past even where meteorological records do not show increasing severity of weather events (see chapters by Glantz, by O'Brien and Liverman, and by Heathcote, this volume). Simply projecting dryland degradation, for example, highlights the need for long-term strategies to stop or reverse these trends in order to improve the productive capacity and security of populations in semi-arid regions. Each year large areas are being at least temporarily worked to the point of declining productivity (Ocana 1991:3; WRI 1991, 1992).

Today, in the semi-arid regions, vulnerability to the consequences of existing climate variation is already a major problem. Dryland degradation is widespread and progressive, while semi-arid populations are growing. These trends only compound the vulnerability of people and of social systems. Without addressing current problems, future vulnerability can only get worse, exacerbated or not by climate change, making the magnitude of future crises even greater. By addressing today's vulnerability we can increase the ability of semi-arid regions to adapt to and cope with the as yet unknown characteristics of a future climate change. Actions taken today to reduce vulnerability – actions which have been justified for a long time – will increase resilience and security by providing a buffer against vulnerability to future consequences of climate change. These are called 'no-regrets policies,' since they are valuable actions regardless of climate change probabilities.

From impacts to vulnerability and beyond

Climate *impact* assessment addresses the magnitude and distribution of the consequences of climate variability and change. *Vulnerability* assessment extends impact assessment by highlighting *who* (as in what geographic or socioeconomic groups) is susceptible, *how* susceptible they are, and *why*. Clearly these assessments are overlapping and interlinked. For informed policy-making purposes, both are necessary and neither is sufficient. Vulnerability analysis ensures that the assessment of impacts will be extended into the realm of social, political and economic causality that shapes susceptibility to impacts. Understanding causality, facilitates appropriate policy design.

Climate impact analysis often focuses on the range of consequences of a given climate event. Examining impacts is a way of looking at the range of consequences of a given stimulus. For instance, drought is associated with a number of outcomes including reduced crop yield, reservoir depletion, hydroelectric interruptions, dryland degradation, and some second-order effects such as economic loss, hunger, famine or dislocation. This type of analysis helps to focus attention on the range of outcomes associated with climate variability or change. But it is somewhat misleading to designate these as *climate* impacts, since they are usually the result of a *multitude* of causal agents. These may include level of development, market organization and prices, entitlement structures, access to productive resources, distribution, state policies, and local or regional conflicts (Blaikie 1985; Watts 1987*a*; Drèze and Sen 1989; Downing 1991, 1992; Schmink 1992). It is some combination of these factors, not the singular result of drought, that makes a family, household, enterprise, nation or region vulnerable. Vulnerability occurs at a conjuncture of physical, social and political-economic processes and events. Hence, complete climate impact analyses must include this multi-causal perspective, placing climate as one causal agent among many.

Downing (1992) presents a method for analyzing vulnerability in which he lays the groundwork for examining this conjuncture in a systematic way. In Downing's framework, vulnerability focuses on *consequences* such as dislocation, that is, vulnerability to having to migrate to the city or to some other frontier. Drought might be considered a cause, even a trigger of outmigration. But, outmigration is also examined as a function of such factors as exploitation, the lack of local alternative income opportunities or high food prices. So, this analysis aims to reveal the range of causes of this outcome – which is of particular social concern – rather than focusing on the impacts of one of many causes or triggering events.

Analysis of vulnerability focuses on the *relative* likelihood of different socioeconomic groups of geographic regions to experiencing each outcome. Hence, relative levels of vulnerability to hunger can be mapped out spatially (see, for example, Box 3, p. 30), temporally and socially. Spatial factors might include location on the rainfall gradient or on a geopolitical map, location with respect to transport or marketing systems, or *vis-à-vis* soil types and other geo-climatological factors. Temporal factors might include coincidence with an economic recession or depression, or perhaps the particular moment in political or development history of a region or country. And socioeconomic factors would include the level of economic development, type of livelihood, level of education, political party or socioeconomic group (gender, class, ethnic group, caste or religion).

By understanding socioeconomic and political factors associated with vulnerability, one can begin to trace out the chains of causal forces and relations that impinge on a given instance of environmentally related vulnerability, chronic deprivation or crisis. In the same way that human vulnerability is shaped by a multitude of causal agents, land degradation, deforestation and other forms of resource degradation are also located within a nested set of causal agents. They too can be evaluated in a similar way (see Blaikie 1985; Blaikie and Brookfield 1987; Schmink 1992). With an understanding of causality, appropriate policy responses can be developed to redress the causes of vulnerability, rather than just responding to its symptoms.

To address each of the causes of vulnerability or environmental decline might require policy interventions at different levels. Political-economic and geographical analysis of vulnerability's causes can be specific enough to allow policies to be tailored for a specific population, place and problem. And finally, since causality can be traced to international, national, household and individual levels, policies can be targeted at the appropriate level if the causes are understood.

In short, the object of vulnerability analysis is to link impact analysis to an understanding of the causes of vulnerability in order to facilitate a meaningful policy process. But, in carrying out such an analysis, one must be extremely careful not to mix correlation with causality. To map out the proximate vulnerability factors, such as location, livelihood, education and income level, tells only part of the story. Without looking at structural causes, such as the way in which the farm economy is embedded within a larger extractive economy, it is difficult to target extractive mechanisms, such as rent structures, sharecropping contracts, usurious credit arrangements, terms of trade and tax structures as causes of vulnerability and environmental decline (de Janvry and Kramer 1979; Deere and de Janvry 1984; Bitoun *et al.*, this volume). Hence, to reduce vulnerability, policy analysts must go beyond identifying its proximate causes to evaluating the multiple causal structures and processes at the individual, household, national and international levels.

We highlight this aspect of climate impact analysis since it (1) allows for a multi-level, multi-sectoral policy analysis, and (2) facilitates the analysis of both proximate causal factors and the broader political-economic forces that shape vulnerability.

Toward ecologically sound development

Access to education, employment, credit, licenses, markets, a healthy environment, land and labor are integral for development. Those on the social and geographical margins need to be able to diversify their income-generating activities in order to reduce their vulnerability. They need an income sufficient to invest in the maintenance of their land and in the stocking of buffers against adverse climatic events, as well as in non-climate-dependent production and survival strategies. The inability of peasant farmers to save and obtain necessary productive resources is a primary structural constraint on their ability to maintain and improve marginal agricultural land. Hence, poor access to infrastructure, inputs, markets, land and credit must be redressed in order to reduce or reverse the rural ecological decline currently under way in much of the semi-arid tropics around the globe. But, given that the processes of differentiation and marginalization that produce the current distribution of assets and patterns of access are ongoing, changes in access must be accompanied by political access to assure that resource access is maintained. They must be accompanied by enfranchisement and inclusion in the political processes (see Drèze and Sen 1989; Watts and Bohle 1993).

An important strategy for relieving a population's pressures on the land and raising rural and urban incomes is to support the development of diverse income-generating opportunities. Diversification of local economies also buffers against severe climatic events. In some regions this may mean fostering existing local productive activities or small-scale enterprises, and in others, encouraging regional pockets of industrialization. Such development is aimed at relieving the local pressures on the resource base and building a buffer against the inherent climatic variability of these regions. But diversification and development will accomplish little if the profits they generate are extracted from the regions and/or concentrated in the hands of a few.

International assistance may be needed for some types of development programs, as well as for avoiding potential ecological problems stemming from development in these regions. In addition, rising greenhouse gas emissions in these regions may need to be offset by reduced emissions or by forest-augmented sequestering elsewhere, such as in the industrial nations of the world. Given the severity of the existing problems the inhabitants and governments of these regions face, they will only be able to address these secondary, less immediate problems of industrial pollution and the emission of greenhouse gases by developing industries with outside assistance. With increased levels of development, the capacity to treat and prevent environmental problems and social vulnerability will increase, and these regions may then move in the direction of more environmentally sound economic-development strategies.

Conclusion

It is important to reduce the emission of gases that are projected to change the world's climate. It is also important to evaluate how that climate change will affect future populations and the future sustainability of the productive natural-

Table 1. *Land area within arid and semi-arid zones in developing regions (%)*

	Africa	Central America	South America	SE Asia	SW Asia, Middle East
Arid and semi-arid lands	66	60	31	33	80
Semi-arid lands	16	22	17	21	12

Source: Adapted from WRI (1990:287).

resource base. But it is equally, if not more important to examine the current environmental degradation and the livelihood insecurity of the vast majority of people living in the world's semi-arid lands. For today's environmental decline will increase tomorrow's vulnerability. Today's vulnerability will reduce tomorrow's resilience. Today's underdevelopment will undermine the potential for increasing future resilience, productivity and development.

There is an old solution to the problems these regions face, and that is development. But this new development effort must occur within the ecological constraints. These constraints are integrally linked to the wellbeing of the most marginal people in these lands. There are numerous technical and institutional measures that can be taken to ameliorate current problems, most of which are worth while even without the specter of global warming – these are the 'no-regrets' strategies. But ultimately, addressing the struggles of the most vulnerable populations in semi-arid areas is what will help them move beyond 'no regrets' to more far-reaching environment and development policies.

The remainder of this chapter is organized into four sections. The first is Semi-arid Regions, in which the characteristics and problems of semi-arid lands are discussed. The second is Climate Variability and Change, which outlines the models and their limitations. The third is Responses, in which some approaches and options for development in semi-arid lands are discussed. A brief conclusion follows.

SEMI-ARID REGIONS

Characteristics

Semi-arid regions are characterized by dry, warm-to-hot extensions of land with low and erratic rainfall, thin and nutrient-poor soils which are prone to salinization, and limited or discontinuous natural vegetation cover (see Fig. 1, depicting the semi-arid regions of the world, and Table 1, showing the proportions of the developing world's land areas

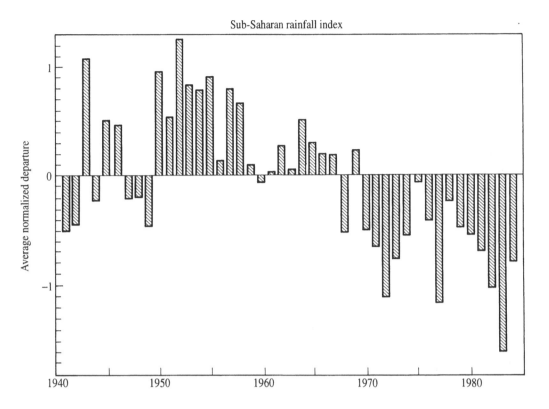

Figure 2: Twenty-year precipitation phases in the Sahel. (Source: Glantz 1989:49.)

that are arid and semi-arid). Drought events, or periods of prolonged dry spells, are inherent characteristics of semi-arid regions. While some semi-arid regions experience season-to-season dry spells, in other semi-arid regions, the dry and wet cycles are much longer, with dry periods extending over a number of years (see, for example, Fig. 2, showing 20-year wet and dry phases in the Sahel). While the vegetation growing in semi-arid regions is well adapted to extreme conditions, the ecosystems are vulnerable to environmental degradation from climatic events and human interventions.

Water is the most precious and limiting factor in agricultural, industrial and urban development in semi-arid lands. Limited water, along with nutrient-poor soils, overgrazing, intensive agricultural practices and mineral extraction, easily surpass the regenerative capacity of the semi-arid ecosystem. They can strip semi-arid ecosystems of their protective vegetation, leading to a process of soil erosion and desiccation, reduced water retention, and ultimately at times desertification (Swindale 1992; Zhao, this volume; Rodrigues *et al*. 1992; UCAR 1991). The term 'desertification' is usually reserved for permanent land degradation. There is, however, considerable debate over the permanence of most land degradation (see the discussion below under Dryland Degradation and Desertification).

Semi-arid regions encompass some of the least-developed areas of the world. Although there is great developmental disparity among the semi-arid regions of the world, they are generally characterized as having a large portion of their population engaged in subsistence agriculture, with a smaller, more economically dynamic sector engaged in large-scale commercial agriculture or industry. The frequent dry periods lead to a high level of uncertainty in production. Those who rely primarily on subsistence plots are particularly susceptible to these variations. Subsistence and pastoral producers frequently migrate within semi-arid regions to more favorable areas, or out of the region altogether, when conditions are sufficiently adverse.

At the regional level, semi-arid regions are on the one hand highly dependent on inputs from and trade with other regions, and on the other are marginalized geo-climatologically, economically, politically and socially. Water resources (as in river headwaters and groundwater recharge sources), agricultural inputs and markets for agricultural products are often centered outside these regions, introducing dependence on the outside, and vulnerability in the face of distribution disruption and market fluctuations. Yet because semi-arid regions have low market-oriented productivity rates (hence contributing relatively little to the gross domestic product), have low population densities, and are typically located far from industrial and urban centers where development policies are generated, they receive low priority in development strategies (Magalhães 1989).

This lack of priority is significant because subsistence and production hardships and opportunities are not based on the physical characteristics of the climate and environment alone. Political, social and economic structures mediate people's ability to cope with extreme climate events and adapt to changing conditions. For example, access to credit, irrigation water, distribution networks, agricultural extension assistance, education and health care are all shaped by people's social, economic and political position in society, and determine their ability to withstand periods of low rainfall (Blaikie 1985; Berry 1989; UCAR 1991:15; Downing 1992). This is particularly significant for the poorest producers, who have limited savings of their own to cushion against failed crops or reduced employment during bad years.

In order to develop policies that can help sustain human activities in semi-arid regions, it is important to examine some of the unique characteristics which have made these regions what they are today: poorly developed economies largely reliant on outside influences and support, susceptible to the effects of their highly variable climates, and at risk of large-scale physical degradation through misuse or overuse. The rest of this section presents a brief discussion of the nature, problems and potentials of semi-arid regions around the world.

Definitions

There are numerous definitions of what comprises a semi-arid region, some more sophisticated than others, but all flawed. The United Nations, for example, has defined semi-arid regions in climatic terms as those areas where the ratio between precipitation and potential evapotranspiration is between 0.21 and 0.50 (P/PET = 0.21–0.50). Semi-arid zones can also be defined as areas where rainfall is between 200 and 800 millimeters per year, and the year-to-year variability is relatively large, at ± 20–30% from the annual mean (Rasmusson 1987). These narrowly technical definitions, however, are problematic on a number of grounds.

First, non-climatic factors such as topography, soil type and cover, and vegetation cover strongly influence local and regional water runoff and retention, affecting the availability of water resources for biological activities, and therefore must be considered in any discussion of semi-arid regions (Yair 1992). The duration and intensity of rainfall have significant effects on whether moisture is absorbed by soils or is lost as surface runoff (Downing 1992). Second, rainfall averages and local hydraulic balances naturally change over time. As a consequence, the physical area that is technically considered semi-arid will tend to shift around, while the reality of a variable climate and a vulnerable ecosystem will remain constant. For example, Sahelian isohyets have migrated 400 kilometers southward since the beginning of the drought in 1968 (Le Houérou 1989:70–4; also see Tucker

et al. 1991). This type of effect is particularly salient where the historical record is short, and there is little basis for determining what is 'normal'. And finally, any discussion of semi-aridity that does not include social components is missing a key set of variables that shape people's ability to produce and thrive in these lands. As Rodrigues *et al.* (1992) points out, the lack of rainfall in an uninhabited region is an uninteresting fact. More comprehensive definitions and approaches can help illuminate the complex environmental and social interactions that shape survival and opportunity in semi-arid regions.

Drought

> Drought implies an extended and significant negative departure in rainfall, relative to the regime around which society has stabilized. (Rasmusson 1987:8)

Oscillating periods of dry and wet weather are a natural feature of semi-arid lands. Some semi-arid regions, such as the Sahel, experience extended dry periods, while others have dry spells during the growing season, or from year to year, as in Canada's Great Plains areas (Glantz 1987; Swindale 1992).

Definitions of droughts, their causes and frequency trends are widely debated (see Box 1, which details some definitional difficulties). Some researchers believe that the severity and longevity of dry episodes witnessed over the past two decades in semi-arid regions around the world represent long-term trends towards greater aridity (Huss-Ashmore 1989:10; IPCC 1990; Zhao, this volume). Some believe that this trend is a natural occurrence, while others attribute it to anthropogenic causes. This latter group points to both global climate change from greenhouse gas buildup, and locally reduced rainfall due to changes in surface reflectance (albedo) and changes in evapotranspiration rates from the clearing or changing vegetation cover (Huss-Ashmore 1989). Other analysts argue that while dryland degradation and global climate change may be occurring, extended drought episodes are in fact the norm in semi-arid regions. Erroneous expectations for consistent wet periods lead to the perception of drought as an anomaly, or as an indicator of worsening conditions (Glantz 1987).

Drought has often been blamed for hardships and lack of development in dry regions. Yet, as Rodrigues *et al.* (1992) points out, 'due to the particular characteristics of semi-arid climates, drought – although not always predictable – will always be a probable phenomenon and thus it should never be considered as a factor of social commotion.' The devastation triggered by drought results from people's *vulnerability* in the face of climatic events, and not the fact of the climatic event itself. As Magalhães (1989:2.27) points out, 'droughts do not cause poverty, they just reveal it, exposing a

Box 1. On defining droughts

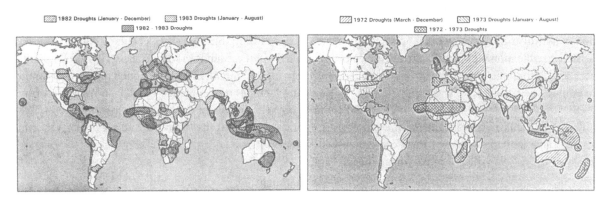

Figure A: Drought in semi-arid regions. (Source: Glantz 1989.)

For the semi-arid regions of the world the single most important, and least welcomed, climatic event is the drought. Drought is an expected phenomenon in the semi-arid regions (see Fig. A). Periods of insufficient rainfall frequently occur within the growing season as well as over entire seasons, making agriculture a risk-prone activity.

The definitional debate on what exactly constitutes a drought is long and as yet unresolved: what may be welcomed as 'abundant' rainfall on the fringes of Africa's Sahel would be accounted as a dry year in North America's corn belt (Hare 1987:4). Yet the debate has led to an in-depth analysis of the concept of droughts and an improved understanding of the climatic as well as socio-economic importance of the phenomenon.

Drought is frequently defined according to disciplinary perspectives, with differing meteorological, climatological, atmospheric, agricultural, hydrologic, water-management, economic and socioeconomic derivatives (Rasmusson 1987; Wilhite and Glantz 1987:14–19; Heathcote, this volume).[1] In discussing the drought phenomenon Wilhite and Glantz (1987) acknowledge that the lack of a precise (and objective) definition of drought in a specific situation has been an obstacle to understanding the concept and has led to indecision and/or inaction on

the part of managers, policy makers and others. Yet they stress that there cannot (and should not) be a universal definition of drought, since definitions should reflect regional specificity. While the available definitions demonstrate a multidisciplinary interest in drought, this interest has not yet translated into a multidisciplinary definition of the phenomenon.

Even though it is recognized that drought is a complex phenomenon with pervasive social ramifications, most scientific research has emphasized its physical, rather than societal aspects. Vulnerability occurs at the conjuncture of drought and a whole host of other physical, social and political-economic processes. Not only are the social and ecological consequences of meteorological drought often caused and aggravated by social and political–economic factors, but these consequences often linger for many years after the event and the secondary and tertiary effects are felt even beyond the spatially defined borders of the drought (Wilhite and Glantz 1987).

[1] The Palmer Drought Severity Index (PDSI), developed in 1965 by W. C. Palmer, is one of the best known and most used definitions of drought. Primarily a meteorological description, it relates drought severity to the accumulated weighted differences between actual precipitation and the precipitation requirement of evapotranspiration (Wilhite and Glantz 1987).

plethora of social and economic problems that remain hidden during normal rainy seasons. When droughts occur, they disturb the tenuous production and survival system of the poor and destroy what little progress in their station was achieved since the previous dry spell.'

Semi-arid regions in developing countries have highly unequal asset distribution patterns resulting from economic and social processes, and government interventions. It is often the result of historical processes, associated with colonization (when previously accessible lands were closed

off through new land-tenure structures) or land 'reforms' in which the most marginal lands were distributed to the poorest producers (Box 2). Government policies, the expansion of large-scale agriculture or livestock production, and population pressures have pushed small farmers into dryer and dryer regions. Where resources are scarce, a period of wet weather can lead farmers to move into marginal regions, made more fertile by the increased rains. When the dry period returns, they are unable to maintain production (see Glantz, this volume).

Box 2. Marginality and drought in Northeast Brazil

Agriculture in the semi-arid areas in Northeast Brazil is predominantly based on small producers, landowners or tenants, and waged workers who produce for their own subsistence, scarcely participating in the market economy. In their paper 'Effects of drought on agricultural sector of Northeast Brazil,' Khan and Campos (1992) focus on these most marginalized groups.

The 1987 'Green Drought' in the Northeastern state of Ceará resulted in a 7% reduction in the area under cultivation. More alarming, however, was the finding that for subsistence crops the difference between harvested production and expected production reached 80% for rice, 75% for beans, and 79% for maize. The earlier 5-year drought (1979–83) had similarly dramatic impacts with the agricultural production diminishing by 83% in relation to 1978, a year with normal rainfall. In each case the scars of drought ran markedly deeper for subsistence farmers, who emerged as the most vulnerable to disaster.

The social result of drought is the formation of virtual clusters of misery and poverty, composed basically of small rural producers who migrate from the countryside to the cities, where they embark upon a futile search for dwindling employment and basic services, adding to the already unmanageable numbers of the displaced. According to a 1984 Northeast Development Superintendency (SUDENE) estimate, the migration balance in Brazil's Northeast for 1980 stood at −5.5 million people.

A study of the 1979 drought showed that the plight of the small landowners and landless producers became all the more miserable because even in times of normal rains their meager resource reserves are barely enough to meet their subsistence needs. Under these conditions, drought acts as an 'aggravating agent' which further depletes their already marginal productivity, leaving them unable or barely able to subsist. It concluded that 'food and debt payment' constituted nearly 80.5% of the 'drought-stricken' worker's expenditure. Debt payment implied goods supplied (primarily foodstuff, plus some other items including kerosene, soap and pharmaceuticals) in advance by the landlord to the workers, sharecroppers and wage earners.

Sometimes, however, these workers succeed in getting temporary jobs at the 'work-fronts' created by the government to provide relief to drought victims by offering employment in government civil works. This also benefits the medium-sized and large landowners by relieving them of having to sustain labor during the drought period. In order to prevent a massive exodus during the droughts occurring between 1979 and 1983, the work-fronts had to create nearly 500 000 jobs in 1979 (9% of the rural economically active population), 720 000 in 1980 (13%), 1.2 million in 1981 (21%), 747 000 in 1982 (13%) and 3.1 million in 1983 (55% of the rural economically active population in 1980). These numbers, however, may be exaggerated primarily because the work-fronts have yet to adopt efficient criteria of hiring, and a large proportion of the unemployed urban populace is also attracted to them. However, the figures highlight both the magnitude of the problem created by labor displacement and an approach towards its solution (Kahn and Campos 1992).

The cycle of drought exacerbates the social inequalities that make dry spells into crises. In times of drought, producers who already exist in precarious conditions are the first to be affected (Watts 1983; Demo 1989; Magalhães 1989, 1992; Khan and Campos 1992; Rodrigues 1992; O'Brien and Liverman, this volume). Pre-existing social inequalities are exacerbated in times of drought, often leading to increased poverty on the one hand and land concentration on the other (Watts 1983; Lemos and Mera 1992:4; Magalhães 1992). Marginal subsistence producers and smallholders have fewer assets to buffer them against poor harvests, and are more likely to resort to selling off assets, eating seeds for survival, intensifying their cultivation practices (thus degrading their soil assets) or abandoning their cultivation altogether and migrating out of the region. Large producers tend to have multiple and diverse assets which they can afford to sell off or utilize more intensively before actually suffering from the effects of droughts. Large cattle ranchers in Northeast Brazil,

for example, who rent land to tenant farmers, will often put that land in forage cultivation to insure that their cattle survive the drought. The tenant farmers are then left with no means of livelihood, and must sell their labor to large farmers – who themselves are cutting their hired labor to reduce operating costs – or migrate to urban centers (Magalhães 1989).

Hence, regardless of the causes of climate variation or change, people's vulnerability in the face of drought is socially shaped. The devastation wrought by recent droughts cannot be attributed solely to climatological events. It occurs largely because nomadic and settled land-use patterns, along with low levels of technical and economic development, have become or been made incompatible with the inherent climatic variability of semi-arid zones. It is clear from this discussion that the effects of drought, and their implications for regional development within semi-arid zones, are socially mediated. Land-holding structures, geographic distribution of holdings, the level of technology, access to resources and markets, and ability to command policy attention are all functions of socioeconomic standing. To address the consequences of drought, people's differentiated vulnerability to its effects must be examined.

Dryland degradation and desertification

Human interventions can increase the productive capacity of semi-arid regions by introducing resources such as water and soil nutrients, thus increasing the intensity of biotic production the system can maintain. This is the ideal. Human interventions can also decrease the productive capacity by reducing soil fertility, water retention capacity and protective ground cover, and causing soil erosion, compaction and salinization. Soils in semi-arid regions, and particularly those in the tropics, suffer significant declines under the pressure of excessive cultivation, livestock herds, and polluting industrial processes. Climatic events such as heavy rains, winds and prolonged dry periods interact with the effects of human actions, exacerbating degradation. If these processes are severe enough, desert-like conditions can result.

Desertification refers to a severe form of dryland degradation. It goes beyond other forms of degradation because its effects may be irreversible. The World Commission on Environment and Development reported that: 'the process of desertification affects almost every region of the globe, but it is most destructive in the drylands of South America, Asia, and Africa . . . Land permanently degraded to desert-like conditions continues to grow . . . Each year, 6 million additional hectares provide no economic return because of the spread of desertification. These trends are expected to continue despite some local improvements' (WCED 1987:127–8). But, like climate change, much remains to be learned about the process of desertification. Indeed, there is no unanimity on the extent or meaning of desertification. Some scientists question the idea of desertification arguing that it has often been based on poor data, research conducted at the end of an exceptionally dry period, and extrapolation of data far beyond the regional specificity of the observed phenomenon (Rhodes 1991; Stevens 1994; Little 1994:1). As Rhodes points out, technical definitions and regional generalizations may do more harm than the attention they focus on the issue does good. But, he continues, 'this assessment should not lead to the conclusion that desertification is not an environmental issue; rather, it should lead to an awareness of the necessity to distinguish between and among dryland degradation processes in order to identify appropriate responses and management strategies' (Rhodes 1991:1141). It can take soils a generation to regenerate, if they can be regenerated at all (Swindale 1992). Thus, whether the creation of desert-like conditions through excessive cultivation

and industrial activities should be called 'desertification' or not, the impact of these activities merits concern.

As Demo (1989:1.14) points out, 'dealing with drought becomes a lost cause for technicians who perceive, sometimes quite competently, the physical restrictions and the reasonable chances of reversing them, but who do not know how to address the political problems.' As with other consequences associated with drought, dryland degradation results from a mix of climatic, social, political and economic conditions and events. The conjuncture of these factors and their human consequences are discussed in the section From Impacts to Vulnerability and Beyond, below.

Regional specificity

No discussion about semi-arid regions can do justice to the conditions – ecological, geological, hydrological, social, economic, or political – that shape development without examining each region in its particular context. Each semi-arid region has its particular characteristics, history, problems and potential. Specific policy recommendations must be founded on the regional and local conditions of each area. As discussed in the following section, general circulation model (GCM) projections, even if accurate, are of little use without a detailed understanding of the local dynamics (see O'Brien and Liverman, this volume; Wang'ati, this volume). This is equally true of the policies that are guided by GCM projections. Indeed, semi-arid zones around the world are quite different, with some highly productive regions in North America and famine-prone regions in Africa. Differences are also reflected in different population densities, production mixes, land distribution, available hydrological resources, technological level, etc. The obstacles and opportunities for development in each region are shaped by these characteristics. Policies to bring about sustainable development must therefore reflect the unique cultural, political, economic and environmental nature of each region.

Diversity in semi-arid lands

The discussion to this point has treated the common characteristics of semi-arid regions, and pointed to some differences among them. There is also tremendous diversity within semi-arid regions. One of the natural development potentials of semi-arid regions is, in fact, the presence of sub-areas within them that have greater natural-resource endowments (Netto et al. 1992; also see Albergel 1992; Cadier et al. 1992; Castellanet 1992; Chevallier et al. 1992; Reyniers 1992; Serpantié 1992). Another resource for development is adjacent resource-rich regions that can help sustain development within the semi-arid regions by providing access to missing inputs such as water for irrigation and domestic use. But such transfers can also be quite dangerous if poorly managed (Zhao, this volume).

Development potential also comes from the diversification of activities within semi-arid zones. While semi-arid regions are largely developed for agriculture and livestock, portions of them may be suited to other activities, such as industry, mining and forestry. Some regional needs can be met using alternative technologies that take advantage of abundant resources in semi-arid zones, such as solar, wind, and sometimes geothermal energy, as is found along East Africa's Rift Valley (Milukas *et al.* 1984). Another way to foster development is to base it on activities that are not vulnerable to the natural climate variations of these regions. Semi-arid regions suffer from structural limitations because of their relative dependence on drought-susceptible agriculture and outside regions for inputs and markets. Industrial, manufacturing, and service-sector growth within these regions would not only reduce climate and market dependence but also increase regional development. Industries that are not dependent on agricultural resources are least vulnerable to climate variation and change. Indeed, there is evidence that industrial growth outstrips agricultural growth in semi-arid zones (World Bank 1990; Magalhães 1992; Zhao, this volume). Diversification of the production mix within semi-arid regions has clear benefits, not only for the development of these regions but also for their ability to withstand the climate variability inherent in them.

In the following section, issues surrounding the modelling and projection of climate change and its impacts are discussed.

CLIMATE VARIABILITY AND CHANGE

Introduction

Climate has always been a dynamic entity. It varies across all terrestrial scales of time and space. Large areas of the earth experience wide uncertainty as part of normal climate. This is especially true of the arid and semi-arid areas, where precipitation varies greatly. Change over longer periods of time is also a 'normal' climatic phenomenon (Riebsame 1989:6).

What makes the current concern for climatic change different from past interest in its perturbations and anomalies is the unprecedented pace and magnitude of the predicted change and the attendant dangers to human and environmental systems. While the global mean surface air temperature has increased by 0.3–0.6 °C over the last 100 years, its average rate of increase during the next century is predicted at 0.3 °C per decade (IPCC 1990:2–3).[3]

This dramatic change is the projected result of increasing atmospheric concentrations of carbon dioxide (CO_2) and other greenhouse gases such as methane, nitrous oxide and chlorofluorocarbons (CFCs). The greenhouse effect is an established physical principle that has enabled life on this planet. It is the accelerated accumulation of these anthropogenic greenhouse gases, however, that threatens to cause rapid climate change (IPCC 1990).

The findings of the Intergovernmental Panel on Climate Change (IPCC 1990:3) mark an emerging, but unsure, consensus amongst experts about the effects of such a greenhouse gas buildup. Attempting to model for the radiative equivalent of a doubling of CO_2 concentration, the major global climate models predict a global-average warming of between 1 and 5 °C (Downing 1992:1). This is not much different from the first such estimate made nearly a century ago, in 1896, by the Swedish scientist Arrhenius. While the apparent similarity of these estimates makes the possibility of global warming appear more likely, the inherent uncertainties of climate modelling science and inadequacies of the modelling tools leave all predictions open to challenge (Stone 1992:34).

In this section we begin by outlining the concept of climate variability and change in semi-arid lands. We then proceed to discuss the limitations of the tools now available for projecting climate change and its consequences. Finally we highlight the urgency of acting within, and despite, the uncertainty of climate-change predictions.

Climate variability

There are two aspects of climate variability that are of concern: its effects on the present populations of semi-arid lands, and the projection of its magnitude and consequences into the future.

Hare (1985:41) defines climatic 'variability' as the observed year-to-year differences in values of specific climatic variables *within* an averaging period (typically 30 years), and climatic change as longer-term changes *between* averaging periods, either in the mean values of climatic variables or in their variability.[4] The distinction between short-term climatic *variability* and long-term climatic *change* is critical. 'One affects the range and frequency of shocks that society absorbs or to which it adjusts, the other alters the resource base' (Parry and Carter 1985:95).

Drought is the most common consequence of current climate variability in semi-arid lands (Wilhite, this volume). And the most vulnerable to its effects are the most marginalized populations: those deprived of the mechanisms and/or resources to prepare for and adapt to climate variation, let alone to climate change (Nobre *et al.* 1992). Ironically, while many recent models and analyses (including the majority presented at ICID) are focusing on the impacts of future climate *change*, the problems of climate *variability*, which may indeed get worse under conditions of climate change, are here today. The consequences are not hypothetical, but are already real and known.

Scientific investigations of global climate change have focused on projecting net or average change, rather than the changed variabilities within it. This focus is due to limitations of the available forecasting tools. Projecting variability is, nonetheless, a major concern in its own right. As an intrinsic characteristic of climate regimes in semi-arid regions, variability defines the many decisions made by those who inhabit these areas (Burton and Cohen 1992). Traditional practices of crop and income diversification, as well as spatial mobility, are a few examples (see Wisner 1976; Parry and Carter 1988; Huss-Ashmore 1989; Watts 1987a). This is an area of research that deserves considerable attention.

Projecting climate change and impacts

Projecting climate change is an important first step in evaluating the consequences associated with global warming. Modelling climate change is inherently difficult. It involves simulating the behavior of intricately linked and complex oceanic and atmospheric processes, some of which are not fully understood. In fact, major scientific uncertainties and knowledge gaps persist at every level from predicting just how fast greenhouse gases might build up to forecasting even the simplest climatic variable of temperature (Stone 1992:34–7).[1]

> Any effort to predict climate changes assumes that climate is predictable – but this is not guaranteed. Forecasts of the efforts of a rise in greenhouse gasses are really just predictions of what will happen in the absence of the unpredictable. (Stone 1992:37)

Working within these uncertainties, and acknowledging them, climate-change projections by the Intergovernmental Panel on Climate Change (IPCC), and others, have used various methods to arrive at best estimates within the available scientific knowledge and tools. Amongst them is the use of historical and paleoclimatic data (IPCC 1990), spatial and temporal analogs (Burton and Cohen 1992) and use of the convenient increment approach (Riebsame 1989). While these have their specific strengths and uses, the most highly developed tool to project climate change is the general circulation model or GCM (IPCC 1990).

Working according to the laws of physics, GCMs simulate possible change by using simplified equations (or 'parameterizations') based in part on current climate conditions and in part on approximations of future factors (IPCC 1990). The projections are, therefore, only as good as the parameterizations. The strength of a parameterization, however, is restricted by two factors: (1) the major gaps in our knowledge and understanding of complex climatic process and systems, and (2) the capacity and speed inadequacies of even the most modern computers that force all climate models to make trade-offs between the number of locations they simulate, the number of climate processes they calculate, and the accuracy of the results (Stone 1992:37).

These general limitations translate into a number of major problems associated with relying too heavily on GCM climate-change projections. The more important include the following:

(1) Different models produce similar trends but differ sufficiently that impact projections can vary with choice of model (Riebsame 1989:66).
(2) Variations in grid sizes[5] and low spatial representation make for coarse resolution of GCMs, and hence local specificity is difficult to obtain (Cohen et al. 1992:11).
(3) Complex topographical features are represented differently in different models with high uncertainty as to how to handle these factors (Cohen et al. 1992; O'Brien and Liverman, this volume).
(4) Sub-grid-scale weather patterns, which can be important determinants of precipitation, are ignored (O'Brien and Liverman, this volume).
(5) Inadequate understanding of and assumptions about cloud formation could result in major errors in GCM results (Stone 1992).
(6) Lack of coupling of GCMs with dynamic ocean and biosphere models reduces accuracy (Downing 1992:2–3).

These problems highlight the dangers of interpolating global projections from GCMs to arrive at regional forecasts. In general, the confidence in regional estimates of critical climatic factors, especially precipitation and soil moisture, is low (see Cohen 1990; IPCC 1990:4; Downing 1992:2; Yair 1992:2; Schmandt and Ward 1992:3; Wang'ati, this volume; O'Brien and Liverman, this volume).

Long-term regional precipitation estimates are even more uncertain than regional temperature projections. This is particularly disturbing in semi-arid areas, where rainfall amounts and patterns are the key variable (Parry and Carter 1988:11). For example, using five major GCMs for projecting climate changes in Mexico, O'Brien and Liverman (this volume) found that the projected changes in annual average precipitation varied from a 23% decline to a 3% increase.

As summarized by Cohen (Fig. 3), the state of understanding in climate-change projections provides credible certainty in the trends in atmospheric compositions; fair certainty in the magnitude of global warming and the regional distribution of its causes; uncertain estimations of the role of global warming, large-scale shifts in precipitation and the magnitude of regional warming; and very uncertain projections on regional water resources. Furthermore, consensus amongst experts is that these uncertainties are not likely to narrow in the immediate future (IPCC 1990) (see Fig. 4 on time scales for narrowing uncertainties).

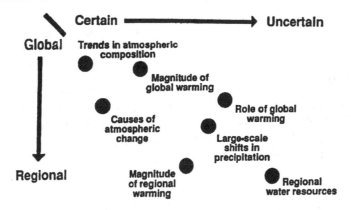

Figure 3: State of understanding of global warming and its consequences. (Source: Cohen 1992).

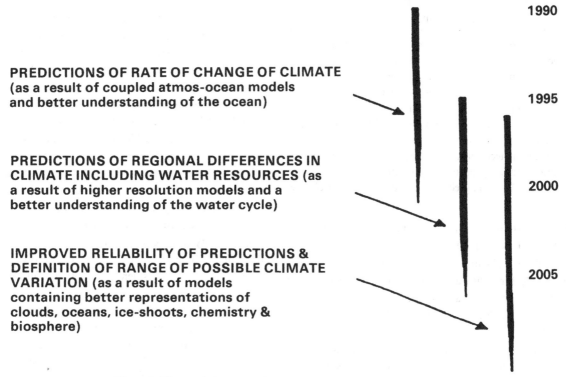

Figure 4: Time scale for narrowing uncertainties. (Source: IPCC 1990).

In the light of all these uncertainties, over-reliance on initial GCM results for projecting impacts and consequences of climate change can often compound issues. This can happen by:

(1) Imposing the climate of the future abruptly on the world of today, without allowing for adjustment and feedback.

(2) Imposing uniform climate changes derived from GCM grid cells onto large regions, thereby eliminating the natural variability in time and space that characterizes real climates.

(3) Dealing with individual economic sectors but not addressing inter-sectoral linkages within the region and/or the linkages between the region and the rest of the world.

(4) Failing to consider how technology and policy may facilitate adaptation to climate change.

(5) Forecasting productivity on incomplete understanding of existing baseline and uncertain projections for changing the technical, social and economic baseline.

(6) Opting for simplistic, and sometimes misleading, carrying-capacity definitions.

(7) Ignoring the role of the socioeconomic structure in affecting vulnerability to climatic change (Rosenberg and Crosson 1992; Downing 1992; Cohen *et al.* 1992; Wang'ati, this volume).

As better computers become available, understanding of physical climatic processes is enhanced, and more developed methods become operational (e.g. fine-resolution, limited-area models) our ability to simulate climate change, and its consequences, is likely to increase, but the ingredient of uncertainty will remain important, at least in the foreseeable future (Cohen 1990, Stone 1992).

Acting in the face of uncertainty

Despite their limitations GCMs are valuable tools for students of climate change. Like any tool, they are most effective when used with care and with understanding of their capabilities and limitations. Climate models are only as good as our understanding of the processes which they describe, and this is far from perfect (IPCC 1990:19).

Despite their many problems GCMs provide basic scenarios against which to explore various climate-change possibilities. Even if we had more refined tools, three crucial problems would arise in projecting future impacts:

> First, of course, no one knows how future climate might evolve; climate forecasting is an uncertain science. Second, any impact projection is only as reliable as the understanding, validity and strength of the assumed relationships between climate and the resource or human activity in question. Finally, even with a good understanding of past and current climate–society linkages, changes in technology and society may exacerbate or mitigate future impacts; and projecting social change is at least as difficult as predicting climate change. (Riebsame 1989:68)

While uncertainty in climate studies needs to be recognized, it must not be allowed to become an excuse for inaction. We should not allow uncertainty to obscure the very real need for policy analysis. Rather, that uncertainty should be incorporated in a credible manner into contemporary policy discussions (Riebsame 1989:68; Burton and Cohen 1992:10).

Many facets of the climate issue, like those of climate variability, are here now. They justify immediate action. The IPCC Working Group on the Formulation of Response Strategies reminds policy makers that amongst the most effective response strategies (especially in the short-run), are the 'no-regrets' strategies – those which are beneficial even without climate change and justifiable in their own right (IPCC 1990:11). Riebsame (1989: 67) articulates the sentiment with much more urgency, pointing out that:

> the large uncertainty surrounding predictions of climate change may not be reduced to levels at which policy makers would be comfortable taking preventive or adaptive actions until the effects themselves become obvious. By then we may be on the verge of unpreventable and irreversible changes in the environment. Uncertainty makes planners adopt a 'wait and see' attitude to account for climate change in their decisions. It can be argued, however, that the changes which might accompany greenhouse warming are sufficiently large and sufficiently immi-

nent (i.e., they will occur in the next few decades) that planners making decisions affecting long-term resource activities such as water development, agriculture, and settlements should consider their implications now.

This general question of action, or response, is discussed in detail in the next section below. While there may be great uncertainties in future climate projections, vulnerability in the face of current climate conditions necessitates and justifies action now.

RESPONSES

Introduction

Current knowledge about climate change is insufficient for planners and policy makers to use models as precise predictive tools. Great uncertainty remains as to the actual mechanisms and potential effects of global warming. While enough is known to argue convincingly that greenhouse gases will result in general climate warming, the specific effects at the regional and local levels are sketchy at best. But it is not necessary to wait for tangible proof of climate change before acting. Climatic variability is currently a major problem in semi-arid regions. In conjunction with other physical, social and political-economic factors, climate variability contributes to vulnerability to economic loss, hunger, famine and dislocation. Reducing this vulnerability by increasing people's ability to cope is an immediate need. This will not only buffer them against the existing climatic variability but would increase their resilience to possible future climate change.

Adaptation is our natural response to the environment. As Burton and Cohen (1992) argue, when and what we plant, how and where we build structures, and what infrastructure we develop, are all contingent responses to environmental constraints. Adaptation is based on immediate observation of our surroundings, our knowledge of environment variation from the past, and on projections of future needs and events. Adaptation is what makes extreme events within the normal climate variation survivable, rather than catastrophic. In highly variable climates such as semi-arid regions adaptive responses mean the difference between a dry spell with some economic losses, and a deadly famine.

No regrets

Climate change introduces the possibility of climatic conditions not previously experienced in a given region. Planners and policy makers are naturally reluctant to invest in adjustments to an uncertain future scenario that may decrease productivity in the short run, or require difficult negotiation with other countries and regions. Yet much adaptation, unlike multilateral agreements about greenhouse gas reduction, is not dependent on cooperation with other nations, can be implemented locally, and can include measures that have immediate benefit, in and of themselves (Burton and Cohen 1992). They are worth doing even if climate does not change. For example, developing drought-resistant crops for semi-arid regions can both help farmers hedge against increasingly arid conditions, and decrease water consumption needed for irrigation (O'Brien and Liverman, this volume). As semi-arid regions become dryer, water conservation will be increasingly important. Using less water is also beneficial because it reduces the likelihood of soil salinization, waterlogging of root areas, and overdraft of the water-table. In regions where water is purchased from sources outside the region, using less water means less expenditure, with resources remaining for reinvestment within the region.

As Wang'ati (this volume) argues, 'sustainable development is achievable provided that development policies put less emphasis on 'change' in favor of 'progressive improvement' on those strategies and technologies which have enabled the populations to cope in the past.' Purposeful adaptation to climate variation will save costs and 'buy time' needed to develop and implement greenhouse gas reduction strategies (Burton and Cohen 1992). The real opportunity in adaptive strategies, however, lies in their coincidence with long-term development aims. Sustainable development does not have to mean sacrificing increased wellbeing in order to preserve resources. Rather, it can mean insuring wellbeing through measures that conserve and improve productive capacity.

Long-term adaptation versus short-term crisis response

> as droughts come and go, left behind are . . . the usual debates over the efficacy of ad hoc relief efforts and at best inadequate or incomplete plans for dealing with future droughts. With the first rains comes a new sense of security, relief efforts are dismantled, plans for the next drought forgotten, and society resumes its so-called harmony with climate until the rains fail and the cycle begins anew. (Easterling 1987, as quoted by Farago 1992)

Just as policy makers are reluctant to act on uncertain information about future climate change, investments to reduce vulnerability to current climate variability have been insufficient. Crisis-induced responses far outweigh preventive measures (Wilhite, this volume). Famine relief, emergency food and seed distribution during droughts, and public works projects to employ affected rural workers are all part of the crisis response mode. With the possible exception of public works projects, crisis responses tend to have immediate palliative results but no lasting effects on the resilient

capacity of the population. In fact, short-term crisis-induced responses can have perverse effects on regional resilience to severe events by making governments and the population complacent in the face of repeated crisis events (Downing 1992; Wilhite, this volume; Glantz, this volume). Rather than a strategy for change, the crisis management approach can be a recipe for maintaining, or worsening, the status quo. For example, providing food and input supports to farmers who are trying to cultivate extremely marginal lands may allow them to weather the current crisis but leaves them susceptible to future events and without incentive to buttress their defenses against vulnerability (Wilhite, this volume).

Wilhite (this volume) argues that agriculturalists must either be equipped to adapt to the natural variability of the lands they are on, or should relocate to more viable areas. Glantz (this volume), however, argues that all viable agricultural tracts are already under cultivation, suggesting that adaptation is the more promising policy response.

Planning for sustainable development

As will be discussed in the following section, the vulnerability of agriculturalists and pastoralists is not solely due to climatic events, but instead comes at the intersection of social, economic, political and cultural factors. Any approach to developing adaptive responses must include all of these spheres. Our discussion is necessarily incomplete, because the specific responses in any area will vary with local conditions and needs, something that would be impossible to do justice to here. Nevertheless, there are a number of general 'design criteria' to guide policy makers, engineers, local entities and the local population in their attempts to respond to the needs of people living in semi-arid regions.

First, responses to climate variability must be adaptive. By adaptive, we refer not only to the spontaneous adaptive responses of local populations, which generate a certain degree of innovation and locally appropriate responses, but also to purposefully adaptive responses on the part of population and government. Many of these responses will need to be corrective measures to increase people's access to productive resources. They will also need to include forward-looking strategies that aim to increase the availability, stability and resilience of infrastructure, markets, institutions, and productive processes.

Second, adaptive responses must be integrative. Experience shows that strictly technical solutions are insufficient to address the vulnerability of people living in semi-arid regions. Economic approaches can address only part of the problem. 'Getting the prices right' cannot protect the poor against volatile prices and declining terms of trade (Bernstein 1979; Sen 1987). Policy measures must not only include a multi-sectoral analysis, but should be formulated through an inclusive process that involves the entire affected population, from small producers to the business community to donors. This will help assure that technologies and policies match the resources and needs of the people concerned.

Third, responses must be incremental and iterative, guiding existing institutional, technical, and socioeconomic structures to a more flexible and appropriate coexistence with semi-arid conditions. The comprehensive project approach can rarely encompass and project the multiple repercussions of policy interventions, not to mention unforeseen future events. An incremental, iterative approach that incorporates active feedback mechanisms to monitor and adjust to emerging events can help overcome these problems.

Finally, there must be active participation of all the parties concerned. Policy formulation, however 'comprehensive' and 'interdisciplinary,' is useless unless it responds to local needs and conditions. Policy interventions include economic and social measures which shape people's participation in economic development by effecting their access to productive resources. These policies, such as how to structure agricultural subsidies, credit systems, technical assistance, etc., have inherent distributional implications, and are therefore political decisions (Demo 1989; Rodrigues *et al.* 1992; de Almeida 1992; O'Brien and Liverman, this volume).

Demo (1989), Rodrigues *et al.* (1992), Vallianatos (1992), Wang'ati (this volume) and others argue that mechanisms to include community organizations, small-scale producers, and other marginalized populations must be developed if the effects of unequal development within semi-arid regions are to be redressed. Pressure from the affected communities, such as rural labor unions, organized community groups, producer cooperatives, etc., is a key element for transforming unequal social and economic arrangements. Together with purposeful policies aimed at increasing producer access to productive resources, this local pressure may be capable of reducing development imbalances within semi-arid regions. Approaches that recognize these groups as legitimate stakeholders, and provide some means for including them in the decision-making process, is critical in redressing marginalization (Demo 1989).

The following discussion provides an analysis of people's vulnerability to the consequences of climate variability, and the technical, governmental, and international responses that this analysis suggests. Again, though this discussion is necessarily incomplete, the aim is to sketch out some of the more important factors that might help shape policy decisions that are compatible with the development aims of semi-arid regions.

From impacts to vulnerability and beyond

The primary problem of semi-arid regions is not climate variability, drought, soil erosion or floods, but more cen-

trally, people's vulnerability to the effects of these events. While droughts, floods and the ecological character of the land are natural phenomena, vulnerability to the effects of environmental change or natural hazards is a social matter. It is not so much the droughts or floods that are alarming, but people's vulnerability to the consequences associated with them: hunger, famine, dislocation from land or livelihood, economic loss, and the loss of ecological assets. Vulnerability comes at the confluence of underdevelopment, social and economic marginality, and the inability to garner sufficient resources to maintain the natural-resource base and to cope with the climatological and ecological instabilities of semi-arid zones.

When we write of vulnerability we are not implying that it is a thing of the future or of mere potentiality. Vulnerability to hunger does not mean that people are not hungry yet. It simply is a way of identifying the populations most likely to be hungry. The word may imply potential problems that are not yet here, but in the semi-arid regions, hunger and frequent crisis are pervasive. We therefore use this term with caution, emphasizing that many vulnerable populations suffer chronic or frequent crises.

Vulnerability is a function of a number of interlinking factors. Neither chronic nor periodic crises emerge from any single agent. Rather, they occur at the conjuncture of many. Famine, for instance, occurs at the intersection of phenomena such as drought, human need, grain prices, wars or frontier settlement policies. It is not a singular result of drought. One aim of this section is to sketch out these forces in order to help develop their policy implications and point to policy responses.

Because the purpose of this chapter is to cover the climate-related issues most pressing in the semi-arid regions of the world, we focus on those groups within these regions whose livelihoods and lives are most at risk. These include the landless and smallholder farmers, pastoralists, and small ranchers in the less-developed regions whose physical well-being is tied to the rains and the land. Risk of economic loss – which in this case is rarely life threatening – is the primary concern within more-developed semi-arid regions. We also examine the linkages between the most vulnerable populations and the population at large, because without a broad and integrated economic development of these lands, chronic underdevelopment and frequent crises will continue.

This section is devoted to defining and discussing the concept of vulnerability.

A formal definition of vulnerability

Here, we adopt Thomas E. Downing's method for evaluating vulnerability when assessing the potential consequences of climate variability and change. Downing's (1992:3–4) methodology consists of:

identifying the multiple dimensions of vulnerability . . . [to a specific consequence such as hunger]; determining socioeconomic groups with similar patterns of vulnerability; assessing their location and degree of vulnerability; delineating pathways by which their vulnerability may be altered by trends in resources (including climate), population, and economy; judging the risk of future climate change, in the context of other expected risks to sustainable agricultural development; and, finally reviewing potential responses that reduce the risk of adverse climate change and enhance the prospects of food [or for our purposes food, job or economic] security. (Downing 1992:3)

Focusing on food insecurity, Downing describes vulnerability as 'an aggregate measure, for a given population or region, for the underlying factors that influence exposure to food shortage and predisposition to its consequences' (Downing 1992:4). Below we discuss the principal characteristics of the concept of vulnerability.

Adverse, specific consequences

The concept of vulnerability is linked to *adverse* consequence. Hence, the concept has an ethical basis – in focusing on the adverse outcomes – which distinguishes it from more neutral terms such as 'sensitivity,' 'consequences' or 'impacts.' Indeed, the concept is designed to help identify those groups within society most likely to experience *negative* outcomes. While crop yield may be sensitive to drought, different households may be more or less vulnerable in the face of the same low-rainfall event. For example, those who have excess grain to sell at a high price during drought-triggered food scarcity will benefit from a drought. Those who have water resources on their private property can sell access or use it to produce goods that are otherwise made scarce by drought. In addition, those with capital can buy land and equipment at low prices from those who are forced to sell out of desperation (see, for example, Demo 1989; Magalhães 1992; Rodrigues *et al.* 1992). While these households are affected by drought, they are not as vulnerable to its consequences as are others in the same community (Downing 1992:5).

Vulnerability is *specific* in that it is concerned with a particular consequence, such as famine, hunger or economic loss. Vulnerability to famine, or to these consequences, can then be evaluated with respect to multiple events such as drought, access to resources, market fluctuations, state policies, or regional conflicts. This is a fundamentally different formulation from previous analyses which link a single cause to an outcome, such as drought to crop yields. Rather than focusing on the consequences of a single event, vulnerability analysis traces out the multiple causes of a single consequence.

The several specific negative consequences that are discussed in the climate variability and climate-change literature include vulnerability to dislocations, hunger, famine and economic loss. These can easily be extended to include the vulnerability to loss (or degradation) of assets, which in turn can be broken down into natural assets such as land, forests and water resources, and human-made capital such as farm machinery or other infrastructure. While these consequences are often discussed together, vulnerability to them must be evaluated separately, for each may have different causes.[6]

Vulnerability is relative

Vulnerability is a scale of the relative likelihood of different socioeconomic groups and geographic regions experiencing negative consequences, such as hunger, famine, economic loss or the loss of productive assets. While everyone is susceptible to all of these adversities, some socioeconomic groups and some areas are more susceptible than others. Clearly the semi-arid rain-fed agricultural regions at the tail end of the rainfall gradient are more likely to experience famines than are cities (see for example, Box 3, and Gasques *et al.* 1992:38).[7] In the same regions, the poor are more likely to experience hunger than are the rich. While all are vulnerable to food shortages, some groups and regions are more vulnerable than others. And, as Downing (1992:4) states, 'ultimately, the analysts must assign the thresholds for concern and action.'[8]

Vulnerability and socioeconomic status

Vulnerability is a function of the relative status of socioeconomic groups. As we will see below, vulnerability is a function of income as well as class, caste, clan, religion, political party, livelihood, race, ethnicity, family, gender and age. Different socioeconomic groups have differing assets as well as differing levels of access to productive resources. Their assets and access are critical aspects of their vulnerability.

Vulnerability is also a function of the degree of development. As Wilhite *et al.* (1987:558) point out:

> In developed countries the proportion of commercial agricultural producers who can withstand a short-term occurrence of drought is high, in terms of both business resilience and human welfare. The impact of short-term drought is significantly reduced because of irrigation and the availability of sufficient forage (fodder) and water for livestock. Even in the case of longer-term drought, the use of irrigation coupled with sufficient grain and forage storage facilities and a fully developed infrastructure can significantly lessen the impact on society and livestock. For example, in Australia, 40% of the farming community is not greatly affected by a drought.

For subsistence farmers, even a short-term drought can be disastrous, especially for the peasant farmer whose only security is a small piece of land on which to grow food and

Box 3. Climate change and vulnerability in Kenya

Table A. *Estimated prevalence of food poverty in 1984 and sensitivity to climatic variations in Kenya*

	Central	Coast	Eastern	Nyanza	Rift Valley	Western	NE	Nairobi	Total	%
Estimated prevalence of food poverty in 1984 (in 1000s)										
Pastoralists										
Nomadic	0	37	77	0	713	0	268	0	1095	19
Agro-pastoral	1	4	3	0	29	0	11	0	48	1
Migrant farm	0	0	14	0	128	0	48	0	190	3
Landless										
Poor	159	88	155	139	310	45	3	0	899	16
Skilled	0	0	0	0	0	0	0	0	0	0
Rural Landholders										
Large farm squatters	53	62	61	68	36	49	0	0	329	6
Smallholder	555	179	637	705	377	511	5	0	2969	53
Gap farms	0	0	0	0	0	0	0	0	0	0
Large farms	0	0	0	0	0	0	0	0	0	0
Urban										
Nairobi	0	0	0	0	0	0	0	32	32	1
Other	9	26	13	14	16	7	1	0	86	2
Total 1984	777	396	960	926	1609	612	336	32	5648	100
Sensitivity to climatic variations										
Current agroclimatic zone suitable for maize (km² × 10³)	13.2	44.7	51.2	14.2	104.7	9.1	7.8	0.7	245.6	42
Decrease with a 10% reduction in length of growing period (%)	0	21	11	0	21	0	16	0	16	35
Decrease with a 20% reduction in length of growing period (%)	0	29	20	0	28	0	20	0	23	32

Source: Downing (1992).

The first step in analyzing vulnerability to hunger and famine is to identify the socioeconomic groups expected to have different patterns of food security or different levels of food poverty. Given these different patterns, different socioeconomic groups would be expected to have different levels of risk in the face of climate change or other resource, social and economic perturbations. In most instances vulnerability to hunger and famine can be correlated with principal means of livelihood, skill level, reliability of income (perhaps as related to tenure), wealth, land-holding size and geographic location (this is a function of natural resources, distance from transportation and marketing infrastructure, as well as other political-economic factors that structure development assistance, and so forth). Rates of food poverty – the proportion of people with insufficient incomes to procure the recommended minimum level of food – were esti-mated by Hunt for 11 socioeconomic groups in eight regions of Kenya (see Table A). Though pastoralists show a higher rate of food poverty, the largest vulnerable group – over half of the food-poor population – consists of smallholder agriculturalists (Downing 1992.)

To examine some potential consequences of climate change, Downing uses a scenario in which the length of the growing period is altered through a combination of temperature and precipitation changes. In Kenya, a 10% decrease in the length of the growing period could result from either an increase in temperature of approximately 1 °C or a decrease in the precipitation of around 10%. Assuming a 10% decrease in growing period, there are major shifts in Kenya's agricultural zones. Growing season changes are shown in Figs. B and C. The threshold of reliable agriculture (60–90 days for maize) would move well into Kenya's highlands as the total area suitable for

Box 3. (*cont.*)

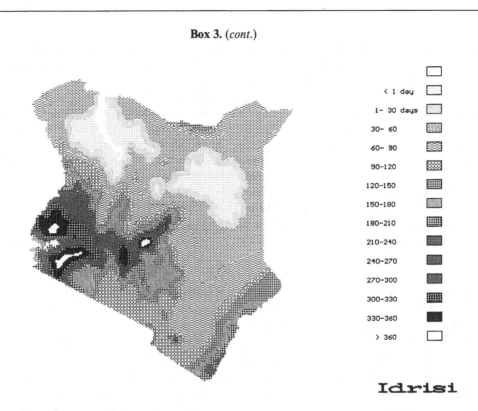

Figure B: Length of the growing period in Kenya (days). (Source: Downing 1992.)

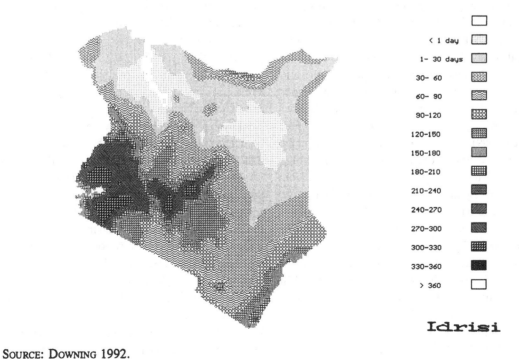

SOURCE: DOWNING 1992.

Figure C: Length of the growing period in Kenya (days) reduced by 10%. (Source: Downing: 1992.)

Box 3. (*cont.*)

maize shrinks about 15% compared with present conditions. While the lowland ranching zone would expand, the tea–dairy zone may disappear, becoming more suitable for coffee.[1]

Downing has estimated the impacts of this climate-change scenario on the socioeconomic groups defined by Hunt; the results are presented in the bottom half of Table A. Clearly, the most direct effects will be felt by groups who rely on their own agricultural production for a major share of their food consumption: pastoralists and smallholder agriculturalists. Reductions of 15–30% in the area suitable for maize cultivation in the sub-humid and semi-arid areas would significantly increase the number of

people with climatic resources inadequate to sustain agriculture. The shortening of the growing season would further increase vulnerability as the probability of achieving yields is reduced.

While other socioeconomic groups will also be affected through changes in the agricultural ecoomy, they may be better able to adapt or adjust to the change. It is those who are already the most vulnerable and most marginalized who stand to lose the most.

[1] A 2 °C reduction, or a 15% drying, or a combination of an increase of 1 °C and a 10% reduction in precipitation, would correspond to a 20% reduction in the length of the growing season, and even more dramatic contractions in the length of the growing season.

---

some cash crops. If seasonal rains fail, no alternative supply of water is available to sustain growth. The result is critical shortages of food, inadequacy of grazing land, suffering and possibly loss of life for both human beings and livestock. The lack of adequate infrastructure (related to the lack of proactive government programs) and the high price of grain from external markets impede governmental ability to rescue inhabitants of such distressed areas.

While China and India used to be thought of as lands of drought and famine, they appear to have reduced their vulnerability by bringing these problems under control – at least temporarily. It is now Africa that is plagued by famines. While the climate variability in India and China has not been altered, critical political, social and economic factors have changed over time. 'Although there were major droughts and climate-induced food shortages around the globe – in various parts of Africa, India, China, Indonesia, Brazil – famines occurred only in Africa' (Glantz 1989:46) (also see Sen 1987; Drèze and Sen 1989). At the same time, droughts in the southwestern United States, central Canada and Australia have led primarily to economic loss (Wilhite and Glantz 1987:20–3; Rosenberg and Crosson 1992; Cohen *et al.* 1992:7–11). Here again there is a critical difference in vulnerability to hunger, famine and economic loss from place to place. Climate variability alone cannot account for vulnerability or outcomes. Both the magnitude and the type of vulnerability are different in different regions of the world, at different times, under different social and political-economic conditions, and different levels of development.

Developing countries are more vulnerable than developed ones. Not only does the level of vulnerability differ, but the type of vulnerability experienced in developing regions differs from that experienced in the industrial countries of the world. Magalhães (1991:7) point out that 'while in developed areas the impacts are mainly of an economic and environmental nature, in developing areas they are mostly social.'

Clearly, vulnerability is a strong function of the level of regional development.

Vulnerability, causality and policy

Analysis of vulnerability has several ramifications in the policy sphere. In giving a *relative* indication of the level of vulnerability, policies can be aimed at the most vulnerable populations. Policy priorities can be established according to need. And, since vulnerability analysis focuses on the *multiple causes* of a single consequence, it allows policies to be designed for the range of causes that make climate events into economic and social crises. In addition, given the specificity of this type of analysis, policies can be tailored for a specific population in a specific place. And lastly, since causality happens at international, national, household and individual levels, policies can be targeted at the appropriate level *if* the causes are understood.

Knowing that vulnerability to dislocation, hunger or famine is a function of geographic location and income is a first step in evaluating vulnerability. But this does not tell us why people end up on submarginal lands or how and why they are impoverished. The causes of such spatial and economic marginality, and hence vulnerability, must usually be understood historically. Such marginality could be partly a function of land concentration, as is the case in Brazil's Northeast (Demo 1989; Magalhães and Glantz 1992; Rodrigues *et al.* 1992). It could be due to state policies encouraging cultivation on marginal lands as in parts of West Africa, or as in Australia it could be a result of people moving onto these lands not knowing that they are making this move during an unusually wet period (Glantz, this volume; Wilhite, this volume). It could also be due to the inability of a farm household to accumulate sufficient capital to invest in the maintenance of the land, or to have sufficient assets to buffer against the consequences of a drought (Sen 1981; Blaikie 1985; Downing 1991; Watts and Bohle 1993). Where these vulnerabilities are a function of class or other forms of social status, they may be results of lack of access to inputs to the productive process. Alternatively, vulnerability may be due to market instabilities or to the classic case of declining producer prices with increasing input and consumption-good prices, as in a 'simple-reproduction squeeze' (Bernstein 1979; Blaikie 1985; Sen 1987; Swift 1989).

The focus on different socioeconomic groups and on specific kinds of vulnerability or threats, such as famine or economic loss, facilitates the policy-analysis process. First, it allows policy makers to identify those groups and regions most at risk. And second, it can illuminate the causal variables, and hence the links to appropriate policy interventions for each group *vis-à-vis* specific kinds of vulnerability (Downing 1992:4).

For example, the analysis of vulnerability to specific outcomes also often reveals different causal factors for different groups. Herders and farmers may be vulnerable to hunger for different reasons. For the pastoralist it may be a function of access to dry season pastures, while for a farmer vulnerability may be due to a low savings rate and a lack of fall-back income opportunities or inability to diversify assets. It is in response to these types of causal factors, and the political-economic forces that shape them, that specific and appropriate policy options can be chosen and applied.

Climate variability affects rich as well as poor sectors of society. But, while hunger, famine and dislocation tend to threaten the poor, economic losses threaten the better off. Those who are well off may experience great material losses without ever going hungry. Thus, policies targeting both food security and economic security may be in order. But accumulation on the part of the wealthy, and policies justifying or structurally disposed to economic growth in already more economically productive regions or on larger farms, have often been part and parcel of the problem of marginalization – marginalization being a flip side of concentration. Hence, special attention must be focused on intervening in ways that do not exacerbate marginalization and vulnerability by reinforcing ongoing differentiation processes. That is, policies must acknowledge the role of accumulation and the lack of it (e.g. differentiation processes) in creating and maintaining vulnerability.

In short, the object of vulnerability analysis is to link impact analysis to the causes of vulnerability in order to facilitate the policy process. But analysts must go beyond the proximate causes of vulnerability to root causes. Correlation does not explain causality. To map out the vulnerability factors or indicators, such as location, livelihood, education and income level, is an incomplete analysis. Without addressing structural causes, such as the political economy of resource access and control – that is, the politics of accumulation and marginalization or the ongoing processes of differentiation – it is difficult to target the root causes of marginality, poverty and the resultant vulnerability to hunger and famine.

It is in response to these non-climatic causes of vulnerability that policies to reduce vulnerability can be made. Because these causes are usually multiple, interlinked and historically contingent, it is all the more important to understand their roots. It is identifying causal links that facilitates a meaningful and effective policy process, for it is in addressing the root causes of vulnerability that vulnerability can be reduced.

Levels of analysis and vulnerability

Causes of vulnerability and environmental decline, and the opportunities for their alleviation, reside at a multitude of levels within the social and political-economic context that shapes the options of individuals, enterprises and farm households. Schmink (1992) uses what she calls a 'socioeconomic matrix' when evaluating the causes of deforestation, and the policy handles by which deforestation can be addressed. This matrix can be adapted to the issues of vulnerability and dryland degradation, as has been done in Table 2. In this schematic, the various forces that bear on vulnerability at different levels are sketched out.

In brief, access to and control over resources necessary for production and reproduction at the local level are shaped by forces at a multitude of levels. Because vulnerability is partly a function of access and control over productive resources, vulnerability itself is shaped by all of these factors. While different levels of the system will exert differing degrees of influence on the local dynamic, all of these levels are relevant. The activities and options of a resource user must be examined from all these levels if a complete understanding of that user's vulnerability is to be achieved.

Conclusion

Climate events, although they can trigger catastrophes or contribute to chronic poverty, do not cause vulnerability. Rather, vulnerability is the product of international, national, household and individual level socioeconomic forces shaping people's livelihood options and choices. Catastrophe, as well as chronic underdevelopment, in the semi-arid zones comes at the intersection of nature and society. It comes at the conjunction of ecological limits, climatological events, the social organization of alternatives available to those pressed by exploitation, market prices, state policies and environmental change against falling productivity.

The most acute problems in semi-arid lands are products of a chronic lack of development, that is, a lack of the resources necessary to hedge against extreme, but expected, events: events that would surprise only a stranger to these

Table 2. *The socioeconomic matrix of vulnerability*

GLOBAL CONTEXT

Markets	*International aid policies*
Demand for natural	Development lending
resources (mineral, forest,	Structural adjustment
and agricultural goods)	Environmental conditionality
Foreign investment	

International agreements and
cooperation
Inter-regional cooperation
Technology transfer
Trade agreements

NATIONAL CONTEXT

Markets	*Policy*
Transportation	Roads and infrastructure
Prices	Price supports and subsidies
Financial markets	Extension services
Migration	*Land tenure*
Population pressures	Land distribution
Frontier expansion	Property regimes

REGIONAL/LOCAL CONTEXT

Settlement patterns	*Interest groups*
Localized population	Conflicts over resources
pressures	Coalition and alliances
Resource distribution	

HOUSEHOLD/COMMUNITY CONTEXT

Gender relations	*Family/community strategies*
Division of labor and	Access to resources
resource access control	Income sources and
Family size and composition	employment
Control over fertility	Temporary migration

Source: Adapted from Schmink (1992).

regions, but which most local farmers know present a risk. It is not that the risk is unknown nor that the methods for coping do not exist, for people have been coping with climate variability for millennia. Rather, inability to cope is due to the lack of – or the systematic alienation from – resources needed to guard against these events.

The central issue of semi-arid regions is one of development and its distribution. Subsistence farmers need a margin of security and a level of savings sufficient to invest in maintaining, upgrading and developing their assets. They need access to infrastructure and to research commensurate with their needs and relevant to their goals. They also need access to alternative income-generating opportunities so as to increase their security in times of drought and to complement their agricultural activities.

In the final analysis, it is accumulation and concentration

that marginalize, and it is marginality that makes people vulnerable in the face of environmental change. Marginality can also push people to 'mine' their natural resources, which only increases their vulnerability. But, so too do accumulation and concentration contribute directly to the ecological decline that makes marginal populations more vulnerable. Certainly, concentration and the drive for accumulation are important causes of widespread ecological insults. Pesticide and fertilizer overuse, uncontrolled effluents and speculative deforestation are all associated with concentration and wealth. These insults all increase vulnerability of marginal populations. They too must be taken into account in evaluating both environmental decline and vulnerability. Indeed, marginality of the environment, just as the marginality of the poor, results in higher vulnerability of both.

Development with a focus on equity and access is part of the solution to this set of problems. Productive security and environmental quality depend on development, access and inclusion – indeed, empowerment and enfranchisement. However, that development must be a conscientious development that accounts for the need to maintain a healthy natural-resource base and a secure population.

Governmental, institutional, and policy responses

Introduction
The problems facing semi-arid regions are multiple and overlapping; so too are the opportunities. There are no purely economic strategies that will reduce regional vulnerability or increase security, just as there are no purely technical strategies that will increase production, reduce poverty and reverse dryland degradation. The papers presented at ICID were unanimous in their call for adopting a multidisciplinary approach to development in semi-arid lands. Any environmentally sound development policy requires an integrated approach that allows for iterative responses to multiple, interrelated spheres – social, economic, technological, biological, political, cultural, etc. – through which policies reverberate and by which policies are reshaped.

The need to reduce and, it is hoped, reverse, the possibility of climate change cannot be overemphasized. Although these responses rightly fall into the government (domestic and international) arena, policies that aim at reducing greenhouse-gas emissions are beyond the scope of this chapter. Rather, we limit ourselves here to governmental, institutional and policy responses that may help enable the inhabitants of semi-arid regions to overcome their vulnerability in the face of current climate variations and possible future climate change. We highlight some of the important ingredients necessary for reducing vulnerability, and increasing resilience to climate variability and change.

Economic development

Environment and development are tightly linked. Unless semi-arid regions are developed, smallholders and landless peasants will continue to exert pressure on the natural-resource base in their attempts to eke out a subsistence living. Due to the lack of access to resources and minimal locally retained surplus, they often cannot buffer against the effects of drought; they are forced to 'mine' the land, simply to survive. Without the resources to invest in the maintenance and development of the land, productivity declines. As alternatives diminish, they are forced to migrate into the cities or onto increasingly marginal lands and new frontiers. Without addressing the lack of adequate farm incomes and alternative income-generating activities, preparing for drought and maintaining the resource base seem to have little chance.

A major question in development debates concerns whether to focus on large-scale and commercial agriculture and industrialization, or to support the small subsistence farmers and pastoralists. The first method has been the mode. The plight of small farmers and the landless has been seen as an inevitable cost of development, as their traditional forms of cultivation become 'outmoded.' It has often been argued that the transient inequities caused by the development process would be offset by the development to come. It would be better, the proponents of this strategy argue, to get the peasant farmers to 'modernize' or to get off the land, into the cities and the non-agricultural sector, or off to the opening of new frontiers. In short, it is better to make room for modern agriculture and industry rather than to support a 'backward' form of subsistence. While there may be truth in the notion that mechanization and modernization of agriculture is inevitable, and in many cases desirable, the dislocation of rural peasants has additional and persistent problems. Cities are overburdened with unemployed, while 'frontier' areas, typically more marginal and fragile environments than those left behind, are stressed by the influx of new migrants. In both places, the rural out-migrants often find themselves in exploitative circumstances similar to those they suffered previously.

Poverty did not disappear with agricultural mechanization. Rather, it increased during the 1960s, 1970s and the 'lost decade' of the 1980s (World Bank 1990), leading to a call for basic needs guarantees. Later, policy makers recognized the environmental ramifications of poverty (Eckholm 1979). Poverty leads smallholders to overuse their resource base, thus increasing their vulnerability in the face of climate variation (Blaikie 1985; Downing 1992). Hence it is in alleviating poverty that people's basic needs can be met, and the resource base maintained. But basic needs remain unmet and the poverty experienced by the majority of the developing world's rural populations persists.

It now seems clear that neither the modernization nor the basic needs development approach is sufficient. Development requires supporting both large commercial agriculture and small subsistence farmers. Dynamic economic regional development relies on well-integrated markets where increased production leads to increased savings, investment and consumption at the local and regional levels. Because semi-arid regions are so dependent on the primary production sector relative to other regions, the cycle of impoverishment of small rural producers and resource depletion described in the previous section signals a structural weakness in the economy (World Bank 1990; Goldsmith and Wilson 1991). This structural weakness is especially significant in semi-arid regions because agriculture there is highly susceptible to wide output variations with small changes in climate variability (UCAR 1991). Unless the pattern can be reversed, it is unlikely that integrated rural development can take root (Goldsmith and Wilson 1991; Bitoun *et al.*, this volume).

Unfortunately, reactive famine relief has long been the only form of policy attention the semi-arid areas have received. These short-term relief measures cannot, however, solve these regions' chronic problems. While responding to emergencies will always be a part of government responsibility, policy makers must focus on developing long-term strategies that reduce semi-arid regions' vulnerability to the consequences of climatic events (Wilhite, this volume). Resilient and dynamic economic development, unlike emergency, stop-gap measures, holds more hope for semi-arid regions' economies, their ecosystems, and the people in them.

Regional development strategies for semi-arid regions often focus on diversifying regional production, particularly through industrialization (Goldsmith and Wilson 1991; Magalhães and Glantz 1992). The hope is that a dynamic economic process will set in, where local businesses and firms will develop to meet industrial input needs, generating an integrated and thriving economy. This strategy is particularly attractive in areas where agriculture is susceptible to wide output variations associated with periodic droughts. Focusing on industries, manufacturing, and product processing industries that are not directly dependent on rain or agricultural inputs can buffer the economy from the vagaries of weather. This strategy, however, is frustrated by the structural limitations of semi-arid regions. The relative dominance of agricultural production, and the dominance of subsistence agriculture in terms of the economically active population, lead to demand structures in semi-arid regions which are highly variable, coinciding with drought events. As a consequence, the expected local multiplier effect of industrial development is hampered, local marketing and repair shop networks do not develop, and it becomes easier for industries and manufacturers to obtain inputs and market products outside the region from more stable sources in more developed areas (Goldsmith and Wilson 1991). The result is well-developed economic linkages with industrial areas and markets outside the semi-arid region, and a disarticulated and stagnant economy within. For the countries of the Sahel, where the entire national territory is often within the semi-arid region, reliance on outside markets for non-agricultural development places them at the mercy of the fluctuations and generally unfavorable terms of international product markets, resulting in large deficits and continual dependence on foreign aid (Watts 1987a; Mackintosh 1990).

This analysis demonstrates the importance of developing a viable local demand structure within semi-arid regions. Without sufficient local demand, no amount of local product diversification will lead to an economic dynamism necessary to absorb labor from the rural sector. At best, a small sector of well-off industries will develop that are less vulnerable to climatic events than the rest of the local economy.

Development strategies must increase agricultural output and income, and absorb labor. Increasing farm output and incomes provides the basis for increased demand structure, and labor-intensive activities absorb under- and unemployed labor (Perrings 1991; Lemos and Mera 1992). As farm incomes increase, consumer demand at the local level increases. Increased demand for consumer products causes local businesses to expand and new ones to form. Part-time workers are employed full-time, and new workers are taken on. As the regional economy diversifies it becomes less susceptible to climatic events. Increased incomes and security allow producers to invest in improving their farm management practices such as mechanization, irrigation and use of improved seeds. These measures often increase the number of crops possible each year, thus increasing year-round agricultural employment and the need for local distributors, repair shops, etc. Hence, the multiplier effect of increased farm output and increased incomes can contribute to integrated and sustainable rural economic development (Ranis and Stewart 1987; Storper 1991; Timmer 1992).

As argued in the section on technological responses in agriculture, government interventions to increase agricultural productivity must be focused on technologies, institutions and infrastructure that are compatible with producers' needs, present knowledge and adaptive capacities. Building on established coping strategies, strategies must be step-wise improvements that increase producers' resilience to climate variation while increasing their ability to produce and save (Rodrigues et al. 1992; Wang'ati, this volume). Clearly, increases in agricultural production and incomes must be accompanied by opportunities to leave the agricultural sector. Income-earning alternatives such as commerce, service, and small-scale or household manufacturing at the local level provide an exit response that can reduce out-migration to major urban centers where industrial and service sectors are insufficient to absorb all the excess rural labor (Perrings 1991; Gomes et al. 1992).

Farm income increases in a highly stratified economy tend to accrue predominantly to the already better-off because access to inputs such as credit, irrigation technologies and water, new seed varieties, etc., is shaped by socioeconomic standing and politics (Demo 1989; Lemos and Mera 1992; O'Brien and Liverman, this volume). Larger producers tend to spend more of their increased income outside the local area, in large cities, while small producers, who are less mobile, consume more locally, contributing more to the local economy. Inputs and markets for commercial agriculture also tend to be located outside the region, particularly in a disarticulated, subsistence economy. A context of skewed land holdings, therefore, will tend to retard any multiplier effects of increased incomes in the rural sector.

These observations lead to two conclusions. The first is that land distribution is a critical issue. The second is that particular emphasis must be placed on increasing smallholder output and incomes (including the creation of job opportunities in other sectors). Large agricultural production and industrial sectors already receive significant government attention and support relative to smallholdings, and generally experience dynamic growth relative to the smallholder agricultural sector. The task is not to focus on poor peasant farmers at the expense of large, and typically more dynamic agriculture and industry. Rather, it is to provide the necessary opportunities to smallholders and landless peasants so that they can enter into a process of dynamic economic growth.

Below we discuss several important factors that need to be considered when structuring policies that will foster dynamic agricultural growth.

Infrastructure

Water storage and irrigation are the first interventions that policy makers think of when trying to improve agricultural output and reduce vulnerability to the consequences of climate variation. Yet building reservoirs is not sufficient. Careful attention must be paid to the ecological ramifications of this approach, to who is going to have access to irrigation water, and to how this access will be controlled. (See the discussion of technological responses, below, for a more detailed discussion of irrigation technologies.) This is particularly important when water infrastructure projects are managed by the private sector, where market-based distribution of benefits will tend to exclude those most in need. Alternative water storage and distribution technologies may be more appropriate to local management practices and social networks (Moench 1991b; Courcier and Sabourin 1992) (see Box 4). Policy makers must assess the viability of multiple-scale and technology alternatives.

In general, irrigated agriculture involves a higher level of

Box 4. Combating drought with small dams

While the *amount* of annual precipitation remains an immediate concern for all semi-arid areas, the *distribution* is equally, if not more important. Many semi-arid regions are characterized by extremely intense, but very short periods of rain. Ironically, this can accentuate the harsh existence of an otherwise dry environment by the ravages of seasonal floods. Much of the precious rainwater can often be lost as runoff (Swindale 1992:7). In other areas, the average annual precipitation may be substantial but the high coefficient of variation perpetuates a continued sense of risk. The Brazilian Northeast, for example, faces severe uncertainty in both the precipitation amount and its seasonal distribution: the annual rainfall between years ranges from 400 to 800 millimeters and the variability coefficient from 0.35 to 0.40 (Cadier *et al.* 1992:1; Rodrigues *et al.* 1992:4).

To combat such uncertainty, water-harvesting structures such as small dams have historically been used in most semi-arid regions of the world. Yair (1992:25) describes how the ancient agriculture system in the Negev desert area in Israel, with annual rainfalls of only 75–100 millimeters, was based on an ingenious water-harvesting system using a network of cisterns.

Recently, as the long-term costs and attendant risks of large irrigation works have become evident, there is renewed emphasis on smaller, less cumbersome and simpler water-harvesting developments such as farm-level ponds, 'dug-outs', check-dams, siltation traps and multipurpose reservoirs. These are seen as ways to boost the farm-level resilience to drought as well as productivity through demand-based water provision (Parry and Carter 1990:163; Moench 1991; Swindale 1992:9).

While discussing the utility of small dams as appropriate on-farm water-management systems for the semi-arid areas Cadier *et al.* (1992:8) describe a manual prepared by the Northeast Development Superintendency (SUDENE) for farmers and pastoralists of the Brazilian Northeast. The manual provides a range of specific and simple guidelines for efficient water harvesting in accordance with varying climatic, terrain and resource limitations. Verifying similar successes elsewhere, the SUDENE manual stresses that 'the advantage of a dam is not necessarily proportional to the stored volume.' Small tanks, when used for multiple purposes – e.g. irrigation, groundwater recharge, 'safety storage' for droughts, aquaculture (in conjunction with poultry, pig or duck farming) – often make for more effective and intensive use of the water than larger ponds.

In India, community-based participatory management of irrigation through siltation traps and check-dams has shown dramatic success with small land-holders. However, this success seems limited to homogeneous, nonstratified, 'tribal' communities, with problems emerging in communities where class, clan, caste and other distinctions create hierarchical orders of access and control over the community resource. Experiments in community management of small-scale irrigation projects are also under way in Africa (Jackelen 1992) (see Box 7). Given the importance of water to farmer communities in semi-arid areas, one can identify both opportunities and dangers in community management of irrigation projects. However, from the initial reports from both India and Africa it seems that small projects are much more conducive to such participatory management than larger ones.

---

technology than subsistence agriculture. The technical 'leap' can often be a significant barrier to small producers. The cost of hybrid seeds, fertilizers, and pesticides must be considered when proposing a transition to irrigated agriculture, and 'packaged' production strategies (de Almeida 1992). Often distribution networks are inadequate to allow the entire package to be applied effectively. While large producers tend to have established relationships with distributors, small producers are often located in remote areas, and may have variable demand for inputs depending on fluctuating household incomes or access to credit, making adoption of high-input technologies difficult. Small producers are likely to apply improvements selectively, and incrementally. Risking an entire crop on what, for the producer, is an untested technology, is an irrational choice. Instead, a small section of a plot may be chosen for a new seed variety, or a small amount of fertilizer will be applied to see the effect (J. Tendler, personal communication 1991). In general, these

considerations indicate that agricultural packages requiring comprehensive application in order to achieve significant results are less appropriate for small producers. Rather, projects should involve a series of incremental improvements that build directly on producer experience, skill and financial ability.

Improved roads, communication networks, rural electrification, marketing networks, and product processing and storage are necessary for making integrated development possible (World Bank 1990; Perrings 1991; Gomes *et al.* 1992; Magalhães and Glantz 1992; Zhao 1992). Whether this kind of infrastructure is put in place during emergency make-work projects during a drought period, or is done gradually as a part of a regional development plan, care must be taken to design infrastructure and prioritized projects so that there the distribution of benefits is as even as possible. For example, wide high-grade roads in a few areas may come at the expense of narrow lower-standard roads distributed over a larger region. Narrower roads will go further, having a more extensive positive effect on the region. These kinds of trade-offs are present in nearly every kind of infrastructure project. A policy that is directed at improving the conditions of the most vulnerable, and fostering an integrated economy, implies that the decision must benefit the greatest number of people possible.

Semi-arid zones are seen as largely lacking in resources (water, soil nutrients, etc.), but they are rich in certain resources. Developing semi-arid regions must take full advantage of these resources. Solar, wind and geothermal energy may provide viable alternatives to dependence on water-based or non-renewable resource-based electricity generation (Wang'ati, this volume; Zhao, this volume). Development of both centralized and decentralized technologies should be pursued, aiming at developing technologies that are easily implemented and not dependent on outside inputs.

Credit, marketing and inputs

Small producers tend to go from season to season with a very limited safety margin because nearly all their earnings and production are used for subsistence or extracted from the local economy. Savings are limited, or non-existent. Lack of savings limits their ability to diversify into alternative income-generating activities, or to invest in increasing the productive capacity of their farms. This does not mean that smallholders are ill equipped to adopt improved management practices and technologies, or to diversify their economic activities. On the contrary, subsistence agriculturalists are by necessity diversifiers, adapters and survivors. Their lack of investment capital is a significant barrier for their adoption of more productive activities. Yet, as we argue, improving the productive capacity of smallholdings and the diversification of local economies within semi-arid regions is a basic requirement for regional development and economic growth. Credit is one possible solution to this barrier. Small producers, however, have less access to formal credit systems than better-off producers. Some have proposed alternative credit institutions that are more accessible to small producers (Caron and Da Silva 1992) (see Box 5). These types of flexible strategies demonstrate how government policies can be formulated to address the needs of small producers, and to structure projects that are responsive to them.

Problems of access to markets and productive inputs can constitute a more significant barrier to economic growth than lack of credit. Government interventions through 'variable' incentives, such as product and input prices, taxes, subsidies, and 'user' charges are one way for governments to intervene in facilitating small-producer access to viable commercialization of their products (Perrings 1991). Prices and financial incentives represent only part of the access equa-

tion, however. Institutional barriers also block small-producer access to productive resources. Perrings (1991) calls for user-enabling measures to correct these problems. These measures might include small grain storage facilities, small-producer product processing and marketing cooperatives, and management structures for common property resources such as grazing areas, water and forests.

Human resource development

Education, health care, safe water supplies and sanitation, and vocational training all fall within the area of human resource development. Projects and programs of these sorts can be grouped into two categories, based on the policy intent. On the one hand, developing human resources means insuring a basic level of wellbeing, related to life expectancy, health, literacy, etc. On the other hand, it means creating the capacity for people to increase their own wellbeing through diversification into more profitable activities, or improving current techniques. The two concepts are clearly related. Without the first, the second is impossible. However, the distinction is helpful in structuring government investment policies. Training that provides skills needed in growing sectors will enable people to move readily into economically viable activities while also contributing to the dynamic growth of the region. Conversely, training for activities which are needed outside of the region can facilitate out-migration when better opportunities are available elsewhere (Wang'ati, this volume).

Agricultural extension and training play a significant role in disseminating new technologies and inputs, responding to producer emergency needs, and developing adapted technologies that reflect the specific soil, water and landholding size agriculturalists are working within. Projects and programs dealing with agricultural research and extension must be oriented towards the needs of small producers, and be appropriate to cultivation limitations inherent in semi-arid regions. Ideally, these programs should develop institutional links with entities at the local level to broaden the dissemination of new adaptive technologies and techniques, and to make local decision makers aware of the need for sustainable and appropriate development policies (Rodrigues et al. 1992).

Alternative income opportunities

Much analysis of semi-arid regions focuses on the pressures of a growing population. This observation is useful only to the extent that it signals a shortage of opportunities available to people living in semi-arid regions. Rather than focusing solely on fertility controls – which are politically charged, difficult to implement, and of questionable efficacy – policies should also focus on improving the options available to people. As mentioned earlier, efforts to equip semi-arid populations with skills that are useful outside agriculture, or

Box 5. Credit for small producers

Small producers, though often viewed as a liability to regional development, can be an engine for growth. Caron and Da Silva (1992) note that by diversifying their production mix, small producers hedge against the impacts of drought, floods and frost. They argue that encouraging this kind of diversification can not only pull small producers out of a cycle of poverty, but stimulate rural growth. By making small-scale commercial activities possible for the rural poor, development programs can help buffer farmers against climate variability and change while stimulating rural dynamism and development with minimum external resources (Caron and Da Silva 1992:8).

Rural families often 'place' family members in alternative activities in order to insure a minimum family income at all times. This may involve: sending a family member to the city to earn wages, raising livestock, introducing technological improvements, wage work on others farms, or producing commercial products for supplemental income in case crops fail. Increasing rural incomes provides a foundation for development of local service sector employment through increased consumer expenditures by farm households. Focusing on small-scale producers is a key element in this growth strategy because smallholders, unlike large rural producers, tend to use labor-intensive technologies, rely on local materials and local craft people for repairs, and spend the majority of their incomes on local goods and services (Ranis and Stewart 1987). Yet small producers are often unable to diversify into profitable activities because they lack capital for investment.

Not only do small producers have little surplus to reinvest, but the formal credit system rarely reaches them (Caron and Da Silva 1992). With no credit history, and little or no collateral, few banks want to lend to small farmers. Further, the small size of the loan, the time needed to process each loan, and the difficulty of lending to often first-time borrowers who may need assistance reading and filling out the forms, all lead lending institutions to shy away from small rural loans (Tendler 1989). Lending institutions that broker small loans for larger banks, such as the Grameen Bank of Bangladesh, can take on these 'difficult' tasks. They provide an example of how credit can work for small rural producers (Caron and Da Silva 1992). By processing loan applications, providing technical assistance for new production activities, these 'banks' have had a significant measure of success where they have been implemented. Because the lending institution works closely with the borrowers, it is able to assess borrowers' ability to repay, independent of collateral and credit history information. The lenders insure repayment by setting up borrowers' groups mutually to enforce repayment among members. Group members only benefit if previous group loans are repaid, thus creating strong social pressure against default.

The alternative credit schemes, however, are not without their problems. One significant issue is the time required for borrowers. The 'solidarity' borrowing group must work cooperatively, or at minimum monitor each other's repayments. This can be a time-consuming activity. People who are excluded from traditional lending institutions also tend to be quite poor. They spend nearly all their time in subsistence activities, and are hardly able to afford this 'extra' time (Boviniv 1986). These programs have been difficult to replicate and expand (Biggs et al. 1990). Much of their success depends on the direct face-to-face contact between the loan broker and the borrower to establish a basis of trust and security that the loan will indeed be repaid. This relationship takes time to develop, or local residents must be discovered who can fill this role. Support for income diversification into small-scale enterprises can also serve simply to 'diversify subsistence' without increasing people's standard of living. The proliferation of similar micro-enterprises increases horizontal competition, reduces profits, and increases labor exploitation (Biggs et al. 1990).

Further, credit is not always the key bottleneck to developing a commercial activity. Often lack of access to markets, inputs and technical training prevents people from starting small enterprises. If these obstacles are not overcome, the chance of default is high (Kilby 1979; Schmitz 1982; Tendler 1989). Policies must therefore be designed with all of these considerations in mind. Research shows that diversifying income generating activities into small-scale enterprises works best when they are located in an agglomeration of complementary firms. The backward and forward linkages among small firms help insure access to inputs and markets, and enable the smaller firms to achieve economies of scale and investment jumps in new technologies that each firm on its own could not manage (Kilby 1979; Peattie 1979; Chen 1986; Schmitz 1990; Storper 1991).

A number of policy interventions emerge from the above discussion. The first involves finding the bottleneck, or missing piece in an otherwise thriving environment of firms within a larger agglomeration of firms providing related goods and services (Kilby 1979; Chen 1986; Tendler 1989). For rural semi-arid regions, this means activities that are tied to the dynamic portions of the economy, typically the non-farm sector (World Bank 1990). Credit is the most common form that the 'missing piece' strategy takes, though technical assistance, marketing and input provision are other examples. Ideally, policy interventions should link services or assist the flow of

---

Box 5. (*cont.*)

inputs and products among existing local firms, allowing idle capacity to be brought into use, rather than trying to create new skills and activities (Kilby 1979). Interventions that permit workers in small-scale enterprises to engage in multiple tasks – whether various income-earning activities, or a combination of income-earning and household activities – allows the enterprise activity to fit into the beneficiaries' multiple needs.

---

even outside the semi-arid regions, is one way to enable people to rise above their bare subsistence condition and to reduce pressures on the land (see, for example, Santibáñez 1992:41). Unless there are income-earning opportunities waiting for them elsewhere, however, there is little hope for such a strategy. It must therefore be linked to deliberate efforts to create new jobs and income-earning activities in semi-arid regions, and to foster the growth of alternative activities such as marketing, intermediary goods production and services. Directing credit for these activities, providing subsidies for start-up enterprises, and creating physical and financial infrastructure will facilitate their emergence.

Emergency responses
There will always be a need for emergency responses to climate events in semi-arid regions (see Wilhite, this volume). No matter how well adapted we are to climate variations, there will always be events which exceed adaptive structures. There are, however, ways to reduce the likelihood that we will be caught by surprise by climate events, or that when severe events occur we are completely unprepared to respond (Wang'ati, this volume). Significant advances have already been made in developing forecasting and early-warning mechanisms (Downing 1992). The task at this point is to increase the link between institutions that have forecasting capabilities and local agencies responsible for responding to emerging crises (Servain *et al.* 1992; Cochonneau and Sechet 1992; Soares *et al.* 1992). Gomes, *et al.* (1992) argue, for example, that the lack of coordination and communication among institutions within the Brazilian semi-arid region leads to ineffectual responses to crisis. This situation must be addressed. While much of the large-scale early-warning information is international in scope, there must be well-established links to the regional and local levels. In this way, not only will semi-arid regions be more prepared to deal with severe climatic events, but scientists and policy makers will learn more about the interrelationships between climate and environment.

Conclusion
This discussion is a cursory review of strategies and approaches that can foster adaptive strategies to the severe climate variability of semi-arid regions. It has not been an exhaustive discussion, but rather an indication of the kind of analytical orientation that is needed in order to address the multifaceted problems facing policy makers at the national, regional and local levels in semi-arid regions. Like any adaptive strategy, policy approaches will themselves evolve as we learn more about the interrelationships between climate and environment, and among the multiple spheres of society – economics, politics, technology, environment and culture. Each region and country must find those strategies that are most suited to its own particular conditions, limitations and capabilities. None of the strategies discussed here are panaceas for the triple burden of underdevelopment, marginalization and climate variation. Instead, they attempt to point to a more hopeful future for these regions, which will only come about through significant effort by policy makers, local entities and the local population.

Technological responses in agriculture

Introduction
Trying to respond to the natural fragility of semi-arid areas by applying unidimensional technical fixes without considering the socioeconomic and cultural context within which they are applied can often lead to disastrous results (Glantz 1989:63). The challenge to policy makers is to show that they are able to develop technical responses to climate variability and change that are sensitive to the social context in which they are to be applied. Fortunately, there is ample reason to believe this can be done. In this section we investigate some of the dominant technological responses that have been applied in response to climate variability in semi-arid areas. Each has strengths as well as limitations, opportunities as well as pitfalls. The difference between understanding or ignoring these is often the difference between success and failure.

Technology has many faces. It ranges from the newest inputs and the most 'high technology' practices, to traditional methods of cultivation. The objective of modern, high technology systems is to maximize profit or yield, using improved crop varieties and livestock breeds, along with fertilizers, herbicides, pesticides and mechanization. The aim of many traditional systems is to minimize year-to-year variation in productivity and, especially, to minimize the risk of total loss (Scott 1976; Popkin 1979; Hyden 1980). The price of this stability is often foregone potential yield. The price of modern, high-yield systems is the lack of a safety net, should the crop fail. Before any technological response can be introduced a thorough understanding of the existing systems, and their internal interactions and dynamics, is necessary. Equally, while estimation of risk levels is an essential component of agriculture analysis, the level of 'risk acceptance' and 'risk affordability' must also be taken into account (Nix 1985:107). Policy makers can learn from traditional coping strategies used by indigenous communities in the search for appropriate technological responses to climate variability.

Below we discuss soil and crop management, water management, high-input agriculture and pastoralism. This is neither an exhaustive list nor an exhaustive critique of the available technological options. The aim is to lay out a broad spectrum of major issues and discuss their relation to climate variability and change, particularly the latter as projected by major general circulation models. In general, any response is only as good as its compatibility with the context – physical, climatological, socioeconomic and cultural – to which it is applied. Its viability depends not on the technology alone, but on how it is managed and where it is used.

Soil and crop management

Soil degradation is probably the most significant threat to sustainable agriculture. Attempts to extend agriculture by expanding agriculture onto marginal soils accelerates the process. Soil erosion, deterioration of soil structure, loss of nutrients or nutrient holding capacity, build-up of salts and toxic elements, waterlogging, acidification, etc., remain constant threats to cultivated tracts in the semi-arid regions. Soil erosion is particularly disturbing due to its high replacement costs.

Soil degradation is largely a function of poor crop management. Yet, even on poor soils (like those often found in semi-arid areas) it can be brought under control. For example, inter-cropping, crop–pasture association, using crops with different root distribution patterns, crop rotations, alley cropping, ridging, terracing, mulching, and zero or minimum tillage check soil degradation while also promoting moisture retention (Watts 1987b:179; Huss-Ashmore 1989:27; Swindale 1992:7). Good soil management is the key to halting dryland degradation, or desertification.

Traditional cropping strategies have evolved sophisticated management practices to maintain soil quality (Huss-Ashmore 1989:28; GOP/IUCN 1991). Increasingly, however, the ability of marginalized farmers and pastoralists to maintain traditional soil and farm management practices is being threatened. For example, the beneficial effects of soil organic matter in both improving soil quality and preventing surface degradation is well known to farmers. As access to the commons decreases or when market prices increase, competing demands for dung and crop residues can trap farmers into practices that they know are not sustainable. Dung, for example, may be burned rather than plowed into the soil as firewood supplies grow scarce or too expensive.

Since plant cover is often sparse in semi-arid regions, evaporation from the soil surface can be a significant portion of the total evapotranspiration. With climate change some models project increased rates of evaporation due to increased temperatures (Rosenberg and Crosson 1992; Cohen et al. 1992; Schmandt and Ward 1992; Downing 1992; O'Brien and Liverman, this volume). Increasing low-water-demand plant cover, either in the form of food and cash crops or in fodder crops and grasses, is an obvious and important response, especially since a simultaneous increase in net precipitation is also projected.

Salinity is a problem for semi-arid lands, especially under warmer climate regimes. But it can be turned to advantage through better management practices if the existing information and experience about saline agriculture is utilized. Headway is already being made in employing salt-tolerant crop and grass species. Fodder crops, in particular, can be grown on fairly poor soils with highly saline water to support livestock (GOP/IUCN 1991).

Water management

Semi-arid areas are characterized by rainfall variabilities both in quantity and in spatial and temporal distribution. Wherever more regular and abundant sources of water are available nearby or where erratic or insufficient rains merit the creation of artificial water reservoirs, irrigation has historically been seen as an ideal response to climate variability. Many irrigation systems have a long record of success; from once being considered famine-prone, the dry plains of China and India are major food producers today. Mexico, Pakistan and Egypt have been equally successful in enhancing their agricultural production through intensive irrigation (GOP/IUCN 1991).

Yet the success of many irrigation systems has not come without its economic, ecological and social price (Swindale 1992:12; Selvarajan and Sinha 1992:2). Salinity, drainage deficiencies, waterlogging, compaction and siltation are just a few of the attendant problems which, if severe enough, become the precursors of desertification. With the expansion of intensive, high-input agriculture – often on soils unprepared for the burden – the problems are compounded (see the section on High-input Agriculture below). While agricultural operations can cause salinity problems even in rainfed areas, salinity is most chronic in irrigated soils. Use of unlined canals, sinking of tube-wells into salt-bearing strata, inadequate drainage and poor water and crop management combine in irrigated tracts to reduce water and soil quality. This can substantially decrease land productivity and seriously affect the quality of drinking water (Swindale 1992:6).

Where irrigation systems or their management are divorced from local social realities they can increase, rather than reduce, vulnerability to the impacts of climatic variations and change (O'Brien and Liverman, this volume). The information in Box 6, which deals with irrigation, shows that it is not irrigation itself, but how and where it is used that aggravates soil degradation. Access and distributional issues of irrigation are germane to all technological interventions. The section on Governmental, Institutional and Policy Responses above, discusses these issues in more detail.

More emphasis and attention has now turned to smaller-scale developments such as farm ponds, 'dug outs,' siltation traps and small multi-purpose reservoirs (Parry and Carter 1990:163; Swindale 1992:9; Wang'ati, this volume) (see Box 4). The aim is to retain the precious rainfall that is so often lost in semi-arid areas due to high runoffs, while improving water-use efficiency. Many of the crop management practices discussed earlier serve the same purposes.

More effective use of groundwater can reduce risk of crop failure in uncertain climates, while also circumventing the need for large and cumbersome water distribution systems (Zhao, this volume). For example, in Mali a pilot project for high-input agriculture with water-pumps is already under way (see Box 7) (Jackelen 1992). The use of dug-wells by Indian farmers for preventing crop losses when temporary rainfall deficits occur, or to recharge soil moisture storage before planting a dry season crop, has contributed substantially to improved farm livelihoods. The use of tube-wells in irrigated plains of Pakistan has similarly given the farmer a solution to waterlogging as well as a source of demand-based water provision. In many desert areas there are vast fossil water supplies which can be exploited for agricultural use (Swindale 1992:9). Groundwater depletion is a risk where recharge rates are slow. Therefore, it is essential to use water as efficiently as possible, employing sprinkler or drip irrigation technologies, or other labor-intensive technologies that achieve the same level of efficiency (GOP/IUCN 1991).

High-input agriculture
The thrust toward intensive agriculture stems not just from technological advances in irrigation systems, fertilizers, seed varieties and hybrid development, but as much, if not more, from population and government policy pressures. With projected change and uncertainty in climatic regimes compounding the already strenuous pressures on the semi-arid zones, high-input agriculture may become a necessary measure for combating vulnerability in some areas (see, for example, Box 7).

Improved cultivars are capable of delivering bumper crops in good years. Examples of the success of high-input agricul-

ture in comparatively well endowed (particularly irrigated) semi-arid lands provides optimism (Selvarajan and Sinha 1992). Yet the need for caution must be emphasized (GOP/IUCN 1991). High-input, mono-crop agriculture lacks the 'safety net' built into traditional systems which use multiple native varieties and rarely experience complete system failure. Further, reliable water supplies, nutrient-rich soils and high technological investments are just a few of the prerequisites for the sustained success of intensive, high-input agriculture. On the more risk-prone rainfed tracts, mounting pressures of production and increasing climate uncertainty could force farmers to adopt intensive agriculture with unsustainable, if not disastrous, results (GOP/IUCN 1991; O'Brien and Liverman, this volume).

Experience points toward homogeneity and the loss of sustainability being conspicuously and directly related. Pests and diseases will always reduce crop yields to some extent, but their effects are more devastating when mixed cropping is replaced by mono-cropping and varieties are replaced by hybrids. The technological response to this technology-driven problem is chemical control (Swindale 1992:3). While pests quickly become immune to pesticides, the latter accumulate and contaminate food and water systems. The use of integrated pest management techniques shows significant potential in resolving this problem (Swindale 1992). Some stability of crop production may be achieved by varying fertilizer applications to offset anomalous climatic conditions (Parry and Carter 1990:161). Since many semi-arid soils are deficient in organic matter and are further degraded by soil erosion, increased fertilizer application is likely to be an important response strategy.

In general, it is now being acknowledged that excessive crop specialization should be avoided, despite the short-term economic benefits it provides. Increasing environmental awareness has stimulated research into alternative agricultural practices, with particular emphasis on learning not only from past follies of intensive agriculture but also from the resilience of traditional systems. The value of crop rotation, multi-cropping, mixed farming, use of cover crops and recycling of agricultural wastes is being recognized (Richards 1985). Most importantly, it is finally being recognized that building upon traditional wisdom and grounding strategies in the realities of specific socioeconomic and natural systems can lead to increased and more stable production potentials (Huss-Ashmore 1989:28; Swindale 1992).

Pastoralism
Le Houérou (1985:155) defines pastoralism as the 'unsettled and non-commercial husbandry of domestic animals' and estimates a pastoralist population of 60 to 70 million people (in 1985), mainly in Africa and Asia. He points out that pastoralism is 'essentially – but not solely – a form of

Box 6. Irrigating semi-arid lands

Buffering farmers in the semi-arid regions from the effects of recurrent droughts and erratic rainfall patterns is crucial to achieving agricultural sustainability (Swindale 1992:10). From historic to contemporary times, irrigation has been a favored response for providing such a buffer. From 137 million hectares (Mha) in 1961, the global irrigated area has increased to 219.7 Mha in 1988 – predominantly in Asia; two-thirds in developing countries; and 91% concentrated in 30 countries. In Asia, 30% of the farms are irrigated; in Africa, the share is less than 2%, although the country with the greatest dependence on irrigation is Egypt (Nemec 1988:217).

Large-scale irrigation works have been the mainstay of drought policy in many semi-arid areas in recent decades (Parry and Carter 1990:163). Irrigation has dramatically increased crop production in many countries (notably India, Pakistan, Mexico, Egypt). Yet it is equally well established that irrigation, when managed improperly and/or when incompatible with the area's ecological, social and economic characteristics, can become a disaster for semi-arid areas and a major anthropogenic cause of desertification (Glantz 1989:66; Rodrigues et al. 1992:7).

Large-scale irrigation works being one of the most intense forms of human interference in the hydrological cycle, can sometimes leave deep scars. Native vegetation and fauna are often the first victims of the huge civil structures required for massive irrigation initiatives. Creation of 'irrigation refugees' (those displaced by the structures of irrigation as well as those whose small land parcels fall prey to the requirements of larger holdings for intensive irrigated agriculture) follows soon after. As denudation and mono-culture result in vermin and pest invasions the immediate use of pesticides accelerates. Mismanaged irrigation works can also lead to a whole host of technical problems pertaining to drainage, waterlogging, loss of water quality, siltation, compaction and salinization. In addition, there are the socioeconomic problems that may arise from subsidized water pricing, unfair access to and control of the resource, marginalization of smallholders, inequities in distribution of benefits, and proliferation of water-borne diseases (Horowitz 1991; Rodrigues et al. 1992:7–9).

Valdemar Rodrigues et al. (1992:9) reports that in the Brazilian Northeast a target of bringing 600 000 ha of semi-arid lands under irrigation (from 261 000 ha in 1980) is now under way in the expectation that irrigated agriculture will provide 20 times the yield of dryland cultivation. However, he quotes the department of drought relief works as reporting that 20% of the area already irrigated is facing problems of salinization, compaction or flooding. J. C. Silva (1988, quoted in Rodrigues et al. 1992:6) estimates this figure to be 30%, reaching 50% in certain areas. As Rodrigues puts it, profits originating from irrigation correspond to the losses from salinization; however, while the profits are whisked away to outside markets, the losses accumulate on the land.

Yet, by balancing these problems, there have been large increases in production of staples such as rice and wheat that have been crucial to reducing food deficits in many developing countries (often in the semi-arid zones) since the early 1970s. In large part, these increases are derived from technological improvements in irrigated agriculture. This has directly reduced vulnerability in the face of drought. The argument, then, is not so much against irrigation, as against the mismanagement of large and/or incompatible irrigation systems (Swindale 1992:12).

This realization, coupled with an increasing acknowledgment of the long-term economic and ecological costs of extensive irrigation works, has already set in; after a peak in the 1970s there is now a marked decrease in the expansion of large irrigation networks (Nemec 1988:218; Swindale 1992:8). The same realization has spurred the need for a renewed emphasis on efforts to improve the efficiency of existing schemes through reduction in seepage and other losses; control of salinization, waterlogging and siltation; improved management of watersheds, water distribution and water use; and greater involvement of farmers and end-users for more equitable and efficient management of systems (Swindale 1992:12).

Fortunately, effective research programs are already under way in several developing countries with dry climates, including India, Pakistan and Egypt (Swindale 1992:8). Substantial information exists on issues such as saline agriculture, groundwater utilization, salt-tolerant species, drainage and no-tillage agriculture. This information promises to lead toward better utilization of existing systems as well as development of more efficient, site-appropriate irrigation technologies for the future.

A recent report on recommendations for drylands research (UCAR 1991:13) points out that although irrigation does reduce the susceptibility of agriculture to the vagaries of climate, in the longer run, irrigated agriculture is limited by the physical environment. Technological advances have not changed these limitations dramatically. In 1878, at the start of scientific management of arid lands, it was reported that less than 3% of Utah (USA) was suitable for irrigation. A 1954 study stated, 'Nobody has corrected that notably since; in 1945 the cultivated area of the state, including dry farms, was 3%' (UCAR 1992:13).

It is in acknowledging and respecting the limitations of the physical, climatological and social environments – and of technological frontiers – that the future and success of irrigation in semi-arid areas depends.

Box 7. Financing pumped irrigation along the Upper Niger River

Located in the Sahel, Mali is one of the poorest countries in the world, with a per capita income of US$ 185 per annum (1988). The Timbuktu region in Northern Mali is characterized by desert and semi-arid terrain, food deficiency, and poor means of transport and communication. Here, along the floodplain of the Upper Niger River where the rural settlements are concentrated, a unique variety of 'floating' rice has been the traditional staple crop; one which has provided predominantly subsistence communities with the ability to cope with their harsh environment.

Sown prior to the flood season, using broadcast methods, this remarkable variety grows with the rising flood tides and is harvested on small boats after the stalks are between 2 to 3 meters high. With yields of between 0.50 and 0.75 tonnes per hectare the indigenous variety does not compare well with other rice species that can produce 4–6 tonnes per hectare in the same area. However, this is more than compensated for by the fact that floating rice requires little labor, no inputs and is grown on extensive floodplains, where land supply is not a constraint.

With the average rainfall decreasing by nearly a third from 215 millimeters (1950–1967) to 145 millimeters (1968–1985), the risk to the system increased as droughts intensified, triggering population dislocations and famine conditions during the late 1970s and 1980s. In his paper 'Economic interventions in the changing environment of a semi-arid zone' Jackelen (1992) describes one approach taken to mitigate the situation.

While the floods were still reliable enough to justify the continued planting of traditional varieties, the decrease in their level from historic averages and the increasing infrequency of the humid surfaces required for planting and seed germination constituted an emergency situation requiring an immediately applicable solution. The United Nations Capital Development Fund (UNCDF) considered that the introduction of small pumps (2–5 horsepower) and sprinkler systems to maintain the use of the floodplain for growing traditional varieties was too untried, complex and cost-ineffective a solution. Instead they decided to intervene by encouraging land- and capital-intensive agriculture using 25 horsepower motor pumps for small-scale (20 hectares) irrigated perimeters and introducing high-input rice varieties.

The choice was based primarily on the known successes of the technology and the fact the enhanced productivity (8–10 times the normal yield) would allow for carrying recurrent costs (both inputs and amortization of pumps) at 20–30% of total production. However, it was recognized that the communities – characterized by low education, stark poverty, hierarchical social structures and little experience in land intensive cultivation – would require

adaptation to high-input agriculture; technical upkeep of motor pumps; financial management for procuring diesel, oil, phosphate, urea and seeds from season to season (amounting in cash and stocks totalling US$ 20 000–40 000) and amortizing the cost of motor pumps (US$ 20 000) over their working life of eight seasons; successfully marketing considerable quantities of paddy; and ensuring a high level of social cooperation, management and participation.

Between 1983 and 1985 eight communities constituting some 2000 families (total population approximately 14 000) were targeted. Intensive negotiations resulted in the creation of informal but functional Village Associations in each community, with elected Executive Committees entrusted with all financial and management responsibilities. On average, each community was required to organize and provide a work force of 400–500 workers for 3–4 days a week during the period of 4 months. Thus, for the overall project, 250 000 work-days were mobilized, two-thirds paid in food from the World Food Program and one-third voluntary.

With all eight communities having had 5 or more years of production, technical assessments indicate that despite the harsh environment, minimal government support and low maintenance facilities (for motor pumps), the introduced technology and practices proved appropriate to the needs and capacities of the communities. The target yield for the project was set at 4 tonnes per hectare to allow all recurring costs to be covered with less than 30% of production. In 1990, only two of the eight communities failed to achieve this target (3.1 and 3.5 tonnes per hectare), two produced substantially above it (6.1 and 5.3 tonnes per hectare), while the remaining four produced around the target (5.0, 4.7, 4.3 and 4.0 tonnes per hectare).

The management of revolving funds by the community has, however, produced mixed results, with half the communities in a poor position to amortize the motor pump costs. Even though inputs for the first 2 years were included in the project cost, the relative magnitude of cash and stocks to be replenished and managed yearly (up to US$ 40 000) was itself a significant factor. In addition, each season required US$ 2500 per motor pump to be placed aside for future replacement (a pro rata calculation per household reveals that the amortizement is roughly equivalent to 125 kilograms of paddy per season). While the communities demonstrated an appreciable understanding of the need to replenish inputs each season, the complex mechanisms for calculation and recovery of the motor pump amortization resulted in an understandable reticence on the part of the members to deliver.

As its most valuable lesson the project highlighted the importance of a simplified, easily understood and easily

---

Box 7. (*cont.*)

operable system for the management of community finances. While affordability was not the only factor in the lack of repayment, it is noted that since production does not represent commercial surpluses and requires a deduction of 20–30% each year, an alternative to straight repayments could be envisioned in the form of loan repayments. This highlights the criticality of focusing on credit financing activities, which could range from handicrafts or livestock forming the basis of micro-lending programs to use of small pumps for traditional agriculture or replacement of motor pumps through credit schemes. Communities that have demonstrated strong social structures for managing and honoring their finan-

cial commitments are, in effect, self-selected candidates for becoming the nucleus of 'community banking' on which to base more complex activities for upgrading social infrastructure and diversifying production.

In addition, it is still unclear whether this type of scheme increases or decreases subsistence vulnerability, since these interventions create dependence on external sources (markets and state agencies) for seed, fertilizer, equipment and fuel. The supply of such inputs is particularly precarious in remote regions, and their affordability can easily be affected by fluctuating prices of either inputs or the resulting crops, while their availability can be disrupted by state policies, conflicts or other events.

---

adaptation of human societies to hazards and hardships induced, and imposed on them, by climatic constraints.'

Recent famines in the Sahel and the subsequent plight of the pastoralists have focused attention on pastoralism, particularly in the context of projected climate change. The present state of crisis is often attributed to anthropogenic impacts such as overstocking, wood-cutting, bush-fires, 'wild' water development, and expansion of intensive cultivation. Le Houérou (1985:173) and Watts (1987a:295), however, describe how herders in Africa utilize complex social relationships as insurance against drought and point out that it is in the strangulation of these social relationships, and not just in the vagaries of climate and population pressures, that the present crisis needs to be understood.

Traditionally, pastoralists have combated the high variability in primary production of rangelands by relying on sturdy drought-resistant breeds, by the mix of livestock held, a sophisticated ethno-scientific understanding of local ranges, and by various sociocultural relationships in addition to the basic nomadic movement (Le Houérou 1985; Watts 1987b). Increasingly, the options to exercise these strategies are being severely curtailed by policies that encourage farm encroachment on pastoral lands, settlement programs and blockages in nomadic routes (Wang'ati, this volume). Even those who simplistically blame overgrazing and overpopulation as the culprits, and feel that de-stocking and resettlement of people is the most viable solution, acknowledge that issues of property rights, access to common property and efficacy of human institutions will frame the success, or failure, of any biological or physical 'fix' to these problems (Swindale 1992:14).

Sedentary pastoralism with particular focus on agropastoral systems provides one response to realizing the production potential of semi-arid areas, providing sustain-

able livelihoods and reducing the need for periodic drought and famine relief (Swindale 1992:15). Such systems would need to integrate traditional knowledge about adaptation of stocking rates, livestock mix and grazing patterns with the introduction of suitable fodder crops, trees and shrubs (including saline-tolerant varieties) and improved techniques such as ripping, subsoiling, scarifying, pitting, contour benching and water spreading (Le Houérou 1985:169; Swindale 1992:15). The viability of such programs would, however, be dependent upon policy changes that support the development of local infrastructure and institutions, and community involvement in projects with particular focus on the displaced nomadic pastoralists.

International responses

Introduction

The global interest accorded to the Brundtland Commission (WCED 1987) and the Intergovernmental Panel on Climate Change (IPCC 1990) testify to the new wave of international concern for the global environment. The expectations surrounding the United Nations Conference on Environment and Development (UNCED) reflect not just the twentieth anniversary of the Stockholm Conference, but more importantly the arrival of environmental issues – particularly global climate change – on the international agenda.

In this section we look briefly at the international concerns for climate change, and what expectations are being associated with the UNCED process. We then discuss why it is important to give priority to the semi-arid regions in international discussions and actions regarding global climate change. We then outline some major facets of required international responses. Finally, we discuss why restructuring international economic interventions is of concern.

International cooperation

The global nature of climate change has triggered global interest and global fears. It has even initiated some fledgling efforts toward global response initiatives; these, however, have been far from satisfactory. Whereas verbal consensus on the need for international response strategies exists, the severe distributional disparities, both in causes and in effects of climate change have made any such effort politically charged. While the overwhelming bulk of additions to the greenhouse gas flux results from the wasteful practices in the North, the countries most vulnerable to global climate change are in the South (Agarwal and Narain 1991). The debate is being framed by the North, for Northern interests, yet the South is more vulnerable, and the North is better able to cope with climate-change impacts. Not only will the South have to bear the effects of climate change, increased climate variability and a sea level rise, and pay part of the price of mitigating effects largely of the North's making (e.g. preserving forests, foregoing CFC use and paying higher energy prices) but there is a real danger of ignoring the North's wasteful consumption habits while placing the blame on the South's population (see Agarwal and Narain 1991).

Global networks and institutions

Despite the pressures for action generated by UNCED, meaningful international agreements to reduce greenhouse gas emissions are still a long way off. There have, however, been a number of important international landmarks in this direction. The Montreal Protocol of 1987 to stop the production of chemicals implicated in stratospheric ozone depletion marked one such landmark – even if limited in scope and coming after a protracted, and somewhat painful, process requiring 'long and difficult negotiations and substantial compromise [on] goals' (Harris 1991:112).

More specifically, no major international policy initiative exists to address the problems of the semi-arid areas in the light of the projected climate change. The need for such an initiative seems obvious. But given the inherent diversity between and within semi-arid regions it must emerge from, and build upon, regional networks. Although institutions such as UNEP or the Food and Agriculture Organization (FAO) have larger mandates, their particular interest in the semi-arid zones can be a potential asset. The Inter-governmental Committee for the Fight Against Drought in the Sahel (Comité Inter-Etat pour la Lutte contre la Sécheresse au Sahel; CILSS) and the Inter-Governmental Authority on Drought and Development (IGADD) in the Sudano-Sahelian region and the International Crops Research Institute for the Semi-Arid Tropics (ICRISAT) are a few examples of institutions focused on issues of semi-arid lands (Swindale 1992; Wang'ati, this volume).

Such global and regional networks can provide a forum for evaluating the vulnerability of populations in semi-arid regions facing climate variability and change, exchanging successful development strategies, and sharing information on climate trends and early-warning techniques. Given the lack of funds and facilities in many semi-arid regions, such networks and institutions have a major responsibility for long-term research at the regional and local levels, with particular focus on forecasting, early-warning, and control or mitigation strategies.

The need for networks also stems as much from the fact that semi-arid regions often encompass many nations – the Sahel spans 22 countries – as it does from the importance of sharing and learning from each other's experiences. At the same time, the need to communicate between the regional and local levels requires a cadre of local experts who understand local as well as regional and international issues and needs. Educating and training a cadre of experts and decision makers may best be accomplished by regional and international networks and institutions.

International economic interventions

International trade imbalances, foreign debt constraints, international commodity markets and pricing structures, the nature of foreign loans and assistance, and the nature and level of investments by multi-national corporations (MNCs) all help shape the internal policies of developing nations. Such influences can often outweigh attempts at environmental and resource management, regardless of how much 'political will' a government has (O'Brien and Liverman, this volume). The era since the Second World War has seen the internationalization of production, finance and services. This often defines policy and policy implications at local, and even village, levels (Watts 1987a:292). Glantz (this volume) points out that international pressures can force countries to shift from food crops to export crops, sometimes causing hunger even where there is no drought.

In response to mounting public concern for the environment, some progress has been made. Efforts to restructure 'assistance' criteria, environmental impact assessments and natural-resource accounting in cost-benefit analysis are a few examples (Meredith 1992a). While all these initiatives are yet too new or untried for a meaningful assessment to be made, their potential merits attention.

Balanced inclusion in the international economy is of particular relevance for the destiny of the developing world and of the semi-arid regions within them. The relative position of weakness *vis-à-vis* MNC investments, international terms of trade and price structures can often force developing countries into decisions to exploit resources. Passed on from the international to the local stage through national policy, such decisions (e.g. growing groundnut for export rather than food crops for domestic use) force the pace of degradation and deprivation on the semi-arid lands.

Why focus on semi-arid lands?

In addressing the threats of global climate change, international attention has, until now, been focused on the issues of coastal and island settlements and on tropical forests. The first (such as the Maldives and Bangladesh) have received attention because they are at great risk of total inundation by rising sea level, the latter because forest management can both reduce the problem and contribute to its solution.[9] Semi-arid tropics deserve more international attention for these same reasons.

Like coastal and island settlements, the semi-arid tropics are at extreme risk in the face of projected global climate change. Rather than inundation by the sea, semi-arid lands risk inundation from increased rains, and desiccation from warming.[10] It is no mistake that the semi-arid southern edge of the Sahara desert is called the Sahel, from the Arabic word for shore or coastline, for the Sahel is the tidal zone between the desert and the humid forests to the south. Here exaggerating or shifting rainfall patterns could make these regions uninhabitable. Increasing desiccation drives the northern limit of cultivation southward, as has happened with past and current droughts, taking vast farming and pastoral regions out of cultivation and pasture, and dislocating the populations who depend on these lands.

In general, the ecosystems in semi-arid regions tend to have poor soils, problematic groundwater resources, and depend heavily on already scant and erratic rainfall; in addition, the spatial marginality of their populations is often compounded by chronic poverty and underdevelopment. This combination makes them extremely vulnerable to the negative outcomes associated with climate change. The semi-arid tropics are in real danger of severe and widespread human and environmental catastrophe (famine, desertification, etc.), and the need for action is urgent.

Degradation of the resource base not only jeopardizes the livelihoods of a region's inhabitants, but the resulting deprivation results in 'spillover' of dislocated populations onto other, often equally fragile, ecosystems. The dislocation of 'ecological refugees' is not a concept unique to Africa. A case in point is the relationship between the Brazilian Northeast and the Amazon, with more than 51% of those migrating into the Amazon reported to originate from the semi-arid Northeast (Bitoun et al., this volume). The emerging pattern is clear: degradation and deprivation in one area can translate into increased pressures on others.

Semi-arid regions are among the least developed regions of the world. Consequently, they are also the most vulnerable to the current consequences of climate variability and the potential consequences of climate change. While mitigation strategies applied to the forest and energy sectors will reduce the potential for future aggravation of an already precarious situation, actions to reduce current vulnerability in semi-arid lands will do the same. Indeed, the object is to reduce or avert human suffering while upgrading productivity from the natural-resource base. Balanced development in these areas has the potential to do both. Equitable development can reduce local and migration-triggered environmental deterioration, as well as vulnerability to hunger, famine and dislocation, by providing the resources necessary to invest in maintenance and improvement of the land while providing jobs to prevent migration into other fragile regions. Clearly, we do not have to look to the future to find risk or need. Development in semi-arid regions will help meet these needs and reduce both current and future vulnerability.

There are opportunities in these regions. These opportunities serve not only development aims, but environmental aims as well. Not only do they stop the exploitative 'mining' of the most marginal lands by the most exploited and marginalized populations, but they also reduce pressure on other regions by reducing outmigration or by encouraging migration into more productive (agricultural and non-agricultural) regions and activities. But most importantly, development measures in these most marginal lands increase the wellbeing and security of those who are chronically exposed to the risk of hunger, famine and loss.

CONCLUSION

It is ironic that we must look into the future, to a time distant enough to be free of commitment to immediate action or change, just to discuss the tragedy taking place before us. We project future climate change and future vulnerability to dislocation, hunger and famine, while vulnerability and crisis are already chronic and widespread. Today, future scenarios allow us to discuss these (otherwise too politically charged) development, environment and equity issues in a public forum. Indeed, they have brought these 'future' issues to the center of international attention. But we must use this opportunity to slide back down the projection lines and point to the crisis at hand.

As illustrated in this chapter, and in the chapters that follow, the semi-arid regions of the world are currently experiencing the insecurity and disruptions that climate-change impact analyses indicate could become more widespread. Addressing the current problems in the semi-arid tropics will diminish the future vulnerability that climate change may exacerbate. Clearly, the long-term future of these and other regions of the world depends on today's quality of life and the sustainability of current practices and social relations within these regions. While it is important to look to the future, it is much more important to act today on what we already know from direct and repeated experience. It is critical to begin investing in the future today, by investing in social changes that can support equitable and ecologically sound development.

ACKNOWLEDGMENTS

We would like to express our appreciation to Thomas E. Downing, Michael H. Glantz, Gustavo Maia Gomes, Oswaldo Massambani, Karen O'Brien, Hermino Ramos de Sousa, Henrique Rattner, Jurgen Schmandt, Bruce Stedman, Judith Tendler and Matthew Turner for the critical insights and feedback they provided in discussions, and their constructive comments on the drafts of this paper. We would also like to thank the staff of the Esquel Brazil Foundation, particularly Elizabeth M. Duarte, Francisco Silveira and Paula Pini, for providing immeasurable support. Our thanks are especially extended to Antonio Rocha Magalhães for inviting us to take on this challenge and for the inspiration that he provided throughout. This work was supported by Esquel Group Foundation.

ENDNOTES

1. Heathcote (1983:16) cites P. Meigs' estimate that semi-arid lands make up approximately 15.8% of the world's land area and the United Nations estimates that these lands cover 13.3% of the world's land. The Meigs estimate is based on those areas in which the ratio of precipitation to evapotranspiration falls between -20 and -40. This corresponds, according to Meigs (published in 1953), approximately to the lands falling between the 200 and 500 millimeter isohyets. The basis of the UN estimate (published in 1977) is not explained. It should be noted that the first estimate might be considered conservative since some definitions currently in use correspond to the lands falling between 200 and 800 millimeter isohyets (Rasmusson 1987). For further discussion of the problems of defining semi-arid regions, see the section on semi-arid regions. The population of semi-arid areas derived from Heathcote's (1983:20–1) figures is approximately 11.25% of the world's total. Heathcote (1983:21) also presents a UN figure (published in 1977) of 10.1% of the world's population living in these regions. The figures for the land areas, and hence the proportion of the world's population living in these areas, are subject to uncertainty and definitional dispute. The figures presented here, given the conservative rainfall range, are probably therefore underestimated.
2. Marginality is a fact, marginalization a process. While the fact of marginality can lead to environmental decline, the process of marginalization should be identified as the cause. Marginality is the result of this process. The consequences of marginality are therefore the consequences of the marginalization process. It is this process that should be examined if marginality and its consequences are to be understood and redressed.
3. The uncertainty range is 0.2–0.5 °C per decade.
4. Emphasis added.
5. The highest-resolution climate GCMs specify the state of the atmosphere at the intersections of a three-dimensional grid. Each grid is divided into sections that are approximately 250 miles on a side and about a mile thick (Stone 1992:37).
6. For an excellent evaluation of the multiple causes of soil erosion see Blaikie (1985) and Blaikie and Brookfield (1987). For a similar analysis of the causes of deforestation see Schmink (1992).
7. Gasques *et al.* (1992:38), for example, argue that in Brazil's Northeast, 'basic food production is concentrated in areas of extreme vulnerability where reduced production from climatic variations is not just an economic question, of reduced output, but rather a question of survival' (translated from the Portuguese by the authors).
8. The analyst, in the authors' opinion, must be thought of as including not only the expert adviser or policy maker, but also the populations being analyzed. Participation and participatory research are discussed in the introduction of the section on Responses.
9. Saving the tropical forests would reduce the release of greenhouse gases (through reduced burning and decay) as well as provide a sink for CO_2 (through forest planting and growth) from other sources, particularly energy.
10. This is not an either/or proposition, but, if rains remain unevenly distributed and drying is intensified, alternating floods and drought could be the mode.

REFERENCES

Agarwal, A. and Narain, S. 1991. *Global Warming in an Unequal World.* New Delhi: Center for Science and the Environment.

Akong'a, J., Downing, T. E., Konjin, N. T., Mungai, D. N., Muturi, H. R. and Potter, H. L. 1988. The effects of climatic variations on agriculture in Central and Eastern Kenya. In *The Impact of Climatic Variations on Agriculture*, Vol. 2, *Assessments in Semi-Arid Regions*, ed. M. Parry, T. Carter and N. T. Konjin, pp. 123–270. Dordrecht: Kluwer.

Albergel, J. 1992. 'Bas-Fonds Programme: suitable technologies for sustained development in a semi-arid zone of West Africa. The case of subterranean dam of Kambo in Mali. Presented ICID, Fortaleza-Ceará, Brazil, 27 January to 1 February 1992.

Bates, R. H. 1981. *Markets and States in Tropical Africa*. Berkeley: University of California Press.

Bates, R. H. 1983. *Essays on the Political Economy of Rural Africa.* Berkeley: University of California Press.

Bernstein, H. 1979. African peasantries: a theoretical framework. *Journal of Peasant Studies* 6:4.

Berry, S. 1989. Social institutions and access to resources. *Africa* 59:1.

Biggs, T. S., Snodgrass, D. R. and Srivastave, P. 1990. *On Minimalist Credit Programs.* Development Discussion Paper No. 331. Harvard Institute for International Development.

Blaikie, P. 1985. *The Political Economy of Soil Erosion in Developing Countries.* London: Longman.

Blaikie, P. and Brookfield, H. C. 1987. *Land Degradation and Society.* London: Methuen.

Boviniv, M. 1986. Projects for women in the Third World: explaining their misbehavior. *World Development* 14:653–64.

Burton, I. and Cohen, S. S. 1992. Adapting to global warming: regional options. Presented at ICID, Fortaleza-Ceará, Brazil, 27 January to 1 February 1992.

Cadier, E., Seraphim, B. and Molle, F. 1992. The small dam: a way to combat drought. Presented at ICID, Fortaleza-Ceará, Brazil, 27 January to 1 February 1992.

Campos-Lopez, E. and Anderson, R. J., (eds.) 1983. *Natural Resources and Development in Arid Regions.* Boulder: Westview Press.

Caron, P. and Da Silva, P. C. G. 1992. Small production and sustained development in the semi-arid tropics: the necessity of a sustainable credit system. Presented at ICID, Fortaleza-Ceará, Brazil, 27 January to 1 February 1992.

Castellanet, C. 1992. Rehabilitation of irrigated perimeters. Presented at ICID, Fortaleza-Ceará, Brazil, 27 January to 1 February 1992.

Chen, M. 1986. Overview of women's program 1980–1986, India and Bangladesh. *Oxfam America*, October 1986.

Chen, R. S. and Parry, M. L. (eds.) 1987. *Policy-Oriented Impact Assessment of Climate Variations*, RR-87–7. Luxembourg: IIASA.

Chevalier, P., Jacques, C., Grouzis, M. and Milleville, P. 1992. Cas d'un éspace Sahélien Africain: La mare d'Oursi, Burkina Faso. Presented at ICID, Fortaleza-Ceará, Brazil, 27 January to 1 February 1992.

Cochonneau, G. and Sechet, P. 1992. Trying to synthesize the annual rainfall measurement of the semi-arid zone of the Brazilian Northeast Region. Presented at ICID, Fortaleza-Ceará, Brazil, 27 January to 1 February 1992.

Cohen, S. J. 1990. Bringing the global warming issue closer to home: the challenge of regional impact studies. *Bulletin of the American Meteorological Society* 71[April]:4.

Cohen, S. S., Masterton, J. and Wheaton, E. E. 1992. Impacts of climate scenarios in the prairie provinces: a case study from Canada. Presented at ICID, Fortaleza-Ceará, Brazil, 27 January to 1 February 1992.

Courcier, R. and Sabourin, E. 1992. Handling of surface waters for the small agriculture in the semi-arid tropics of the Brazilian Northeast. Presented at ICID, Fortaleza-Ceará, Brazil, 27 January to 1 February 1992.

de Almeida, M. G. 1992. Irrigation policy: promises of prosperity in the backlands – the semi-arid region of the state of Sergipe, Brazil. Presented at ICID, Fortaleza-Ceará, Brazil, 27 January to 1 February 1992.

Deere, C. D. and de Janvry, A. 1984. A conceptual framework for the empirical analysis of peasants, pp. 601–611, Giannini Foundation Paper No. 543.

de Janvry, A. 1981. *The Agrarian Question and Reformism in Latin America.* Baltimore: Johns Hopkins University Press.

de Janvry, A. and Kramer, F. 1979. The limits of unequal exchange. *Review of Radical Political Economics* 11[winter]:4.

Demo, P. 1989. Methodological contributions towards societal drought-management practices. In *Socioeconomic Impacts of Climatic Variations and Policy Responses in Brazil*, ed. A. R. Magalhães. Fortaleza-Ceará, Brazil: UNEP.

Downing, T. E. 1991. *Assessing Socioeconomic Vulnerability to Famine: Frameworks, Concepts, and Applications.* FEWS Working Paper, AID Famine Early Warning System.

Downing, T. E. 1992. Vulnerability and global environmental change in the semi-arid tropics: modelling regional and household agricultural impacts and responses. Presented at ICID, Fortaleza-Ceará, Brazil, 27 January to 1 February 1992.

Drèze, J. and Sen, A. 1989, *Hunger and Public Action.* Oxford: Clarendon Press.

Eckholm, E. P. 1975. The other energy crisis: firewood. *Worldwatch Paper 1.*

Eckholm, E. P. 1979. *The Dispossessed of the Earth: Land Reform and Sustainable Development.* Washington, DC: Worldwatch Institute.

El-Shahawy, M. A. 1992. Some impacts of regional warming. Presented at ICID, Fortaleza-Ceará, Brazil, 27 January to 1 February 1992.

Farago, T. 1992. Climatic variability, impacts and sustainability: generalization from regional climate impact studies. Presented at ICID, Fortaleza-Ceará, Brazil, 27 January to 1 February 1992.

Gasques, J. G., (coordinator), C. H. Motta Coelho, M. Bosco de Almeida, F. A. Soares, L. A. C. da Silva, M. J. Nogueira, R. C. Lins, J. G. B. Oliveira, C. J. C. Lins and F. Barreto. 1992. A case study: overall scenario. Presented at ICID, Fortaleza-Ceará, Brazil, 27 January to 1 February 1992.

Glantz, M. H. 1987. *Drought and Hunger in Africa.* Cambridge: Cambridge University Press.

Glantz, M. H. 1989. Drought, famine, and the seasons in sub-Saharan Africa. In *African Food Systems in Crisis*, part 1, ed. R. Huss-Ashmore and S. H. Katz, pp. 45–71. New York: Gordon and Breach.

Goldemberg, J. 1992. The road to UNCED: environment and development. Keynote Address at ICID, Fortaleza-Ceará, Brazil, 27 January to 1 February 1992.

Goldsmith, W. W. and Wilson, R. 1991. Poverty and distorted industrialization in the Brazilian Northeast. *World Development* 19:435–55.

Gomes, E., Da Silva, J. E. and Brandao, M. H. M. 1992. The integrated institutional aspect, basic element of sustainable development: the examples of the semi-arid of the Northeast of Brazil. Presented at ICID, Fortaleza-Ceará, Brazil, 27 January to 1 February 1992.

GOP/IUCN 1991. *Pakistan National Conservation Strategy.* Government of Pakistan/International Union for the Conservation of Nature and Natural Resources.

Hare, F. K. 1985. Climatic variability and change. In *Climate Impact Assessment: Studies of the Interaction of Climate and Society*, ed. R.

W. Kates, J. H. Ausubel and M. Berberian, pp. 37–68. Chichester: Wiley Press.

Hare, F. K. 1987. Drought and desiccation: twin hazards of a variable climate. In *Planning for Drought: Toward a Reduction of Social Vulnerability*, ed. D. A. Wilhite and W. E. Easterling with D. A. Wood. pp. 3–9. Boulder: Westview Press.

Harris, J. M. 1991. Global institutions and ecological crisis. *World Development* 19:1.

Heathcote, R. L. 1983. *The Arid Lands: Their Use and Abuse.* London: Longman.

Horowitz, M. M. 1991. Victims upstream and down. *Journal of Refugee Studies* 4:2.

Huss-Ashmore, R. 1989. Perspectives on the African food crisis. In *African Food Systems in Crisis*, ed. R. Huss-Ashmore and S. H. Katz, pp. 3–42. New York: Gordon and Breach.

Hyden, G. 1980. *Beyond Ujamma in Tanzania: Underdevelopment and an Uncaptured Peasantry.* Berkeley: University of California Press.

IPCC. 1990. *Report of the Intergovernmental Panel on Climate Change.* Geneva and Nairobi: WMO/UNEP.

Izrael, J. A. 1992. Climatic variability and climatic change and the related social, economic and environmental impacts. Presented at ICID, Fortaleza-Ceará, Brazil, 27 January to 1 February 1992.

Jackelen, H. R. 1992. Economic interventions in the changing environment of a semi-arid zone: introducing small-scale irrigation in the Timbuktu region of the north of Mali and assessing community participation based of financial and production indicators. Presented at ICID, Fortaleza-Ceará, Brazil, 27 January to 1 February 1992.

Khan, A. S. and Campos, R. T. 1992. Effects of drought on agricultural sector of Northeast Brazil. Presented at ICID, Fortaleza-Ceará, Brazil, 27 January to 1 February 1992.

Kilby, P. 1979. Evaluating technical assistance. *World Development* 7:309–23.

Le Houérou, H. N. 1985. Biological impacts: pastoralism. In *Climate Impact Assessment: Studies of the Interaction of Climate and Society*, ed. R. W. Kates, J. H. Ausubel and M. Berberian, pp. 155–85. Chichester: Wiley.

Le Houérou, H. N. 1989. *The Grazing Land Ecosystems of the African Sahel.* Berlin: Springer-Verlag.

Lemos, J. de J. S. and Mera, R. D. M. 1992. Rural poverty and sustainable development for the state of Ceará. Presented at ICID, Fortaleza-Ceará, Brazil, 27 January to 1 February 1992.

Little, P. D. 1994. The social context of land degradation (desertification) in dry regions. In *Population and Environment: Rethinking the Debate*, ed. L. Arizpe, A. Palloni and P. Stone. London: Boulder, Colorado: Westview Press.

Mackintosh, M. 1990. Abstract markets and real needs. In *The Food Question: Profits Versus People?*, ed. H. Bernstein, B. Crow, M. Mackintosh and C. Martin, pp. 43–53. New York: Monthly Review Press.

Magalhães, A. R. 1989. Understanding the implications of global warming in developing regions: the case of Northeast Brazil. SQ315-Bl. A–Apt. 104, 70384-Brasilia, DF, Brazil.

Magalhães, A. R. 1992. Understanding the implications of global warming in developing regions: the case of Northeast Brazil. In *The Regions and Global Warming: Impacts and Response Strategies*, ed. J. Schmandt and J. Clarkson, pp. 237–56. New York: Oxford University Press.

Magalhães, A. R. and Glantz, M. H. 1992. *Socioeconomic Impacts of Climate Variations and Policy Responses in Brazil.* Brasilia: UNEP/SEPLAN/Esquel.

Martine, G. 1992. Social development and liberalism: relevant issues for the Brazilian semi-arid region. Presented at ICID, Fortaleza-Ceará, Brazil, 27 January to 1 February 1992.

Meredith, T. 1992a. Adjusting to environmental change impact assessment as a cultural adaptation process. Presented at ICID, Fortaleza-Ceará, Brazil, 27 January to 1 February 1992.

Meredith, T. 1992b. Environmental impact assessment, cultural diversity and sustainable rural development. *Environment Impact Assessment Review* 12:1–2.

Milukas, M., Ribot, J. and Maxson, P. 1984. *Djibouti Energy Initiatives: National Energy Assessment*, VITA/USAID. Project no. 603–0013.

Moench, M. 1991a. *Sustainability, Efficiency, and Equity in Groundwater Development: Issues in India and Comparisons with the Western U.S.* Pacific Institute for Studies in Development, Environment, and Security.

Moench, M. 1991b. *Drawing Down the Buffer: Upcoming Ground Water Management Issues in India*. Pacific Institute for Studies in Development, Environment, and Security.

Nemec, J. 1988. Implications of a changing atmosphere on water resources. Theme paper in *Conference Proceedings of the Changing Atmosphere: Implications for Global Security*. Toronto, Canada, 27–30 June 1988.

Netto. A. V. de Mello, Lins, R. C. and Coutinho, S. F. S. 1992. Exception areas, humid and subhumid, seated in the Brazilian Northeastern semi-arid. Presented at ICID, 27 January to 1 February 1992.

Nix, H. A. 1985. Biological impacts: agriculture. In *Climate Impact Assessment: Studies of the Interaction of Climate and Society*, ed. R. W. Kates, J. H. Ausubel and M. Berberian, pp. 105–30. Chichester: Wiley.

Nobre, C. A., Massambani, O. and Liu, W. T.-H. 1992. Climatic variability in the semi-arid region of Brazil and drought monitoring from satellite. Presented at ICID, Fortaleza-Ceará, Brazil, 27 January to 1 February 1992.

Ocana, C. L. 1991. *Assessing the Risk of Dryland Degradation: A Guide for National and Regional Planners*. World Resources Institute.

Parry, M. L. and Carter, T. R. 1985. The effect of climatic variations on agricultural risk. In *The Sensitivity of Natural Ecosystems and Agriculture to Climate Change*, ed. M. L. Parry, pp. 95–110. Dordrecht: Kluwer.

Parry, M. L. and Carter, T. R. 1988. An assessment of effects of climatic variations on agriculture. In *The Impact of Climatic Variations on Agriculture*, Vol. 2, *Assessments in Semi-arid Regions*, ed. M. L. Parry, T. R. Carter and N. T. Konijn. Dordrecht: Kluwer.

Parry, M. L. and Carter, T. R. 1990. Some strategies of response in agriculture to changes of climate. In *Climate and Development*, ed. H. J. Karpe, D. Otten and S. C. Trindade, pp. 152–72. Berlin: Springer-Verlag.

Peattie, L. 1979. The organization of marginals. *Comparative Urban Research* 7:2.

Perrings, C. 1991. Incentives for the ecologically sustainable use of human and natural resources in the drylands of sub-Saharan Africa: a review. Working paper presented at Technology and Employment Programme, World Employment Programme Research, International Labor Organization.

Popkin, S. L. 1979. *The Rational Peasant: The Political Economy of Rural Society in Vietnam*. Berkeley: University of California Press.

Ranis, G. and Stewart, F. 1987. Rural linkages in the Philippines and Taiwan. In *Macro-Policies for Appropriate Technology*, ed. F. Stewart, pp. 140–91. Boulder: Westview Press.

Rasmusson, E. M. 1987. Global climate change and variability: effects on drought and desertification in Africa. In *Drought and Hunger in Africa*, ed. M. H. Glantz, pp. 3–22. Cambridge: Cambridge University Press.

Reyniers, F. N. 1992. Sketch of the Sudano-Sahelian grain water system that optimizes rainfalls. Presented at ICID, Fortaleza-Ceará, Brazil, 27 January to 1 February 1992.

Rhodes, S. L. 1991. Rethinking desertification: what do we know and what have we learnt? *World Development* 19:9.

Richards, P. 1985. *Indigenous Agricultural Revolution: Ecology and Food Production in West Africa*. Boulder: Westview Press.

Riebsame, W. E. 1989. *Assessing the Social Implications of Climate Fluctuations: A Guide to Climate Impact Studies*. Nairobi: United Nations Environment Programme.

Rodrigues, V. with H. Matallo jr., M. C. Linhares, A. L. Costa de Oliveira Galvão and A. de Souza Gorgonio. 1992. Evaluation of desertification scene in the Northeast of Brazil: diagnoses and prospects. Presented at ICID, Fortaleza-Ceará, Brazil, 27 January to 1 February 1992.

Rosenberg, N. J. and Crosson, P. R. 1992. Understanding regional scale impacts of climate change and climate variability: application to a region in North America with climates ranging from semi-arid to humid. Presented at ICID, Fortaleza-Ceará, Brazil, 27 January to 1 February 1992.

Santibáñez, F. 1992. Impact on agriculture due to climatic change and variability in South America. Presented at ICID, Fortaleza-Ceará, Brazil, 27 January to 1 February 1992.

Schmandt, J. and Ward, G. H. 1992. Climate change and water resources in Texas. Presented at ICID, Fortaleza-Ceará, Brazil, 27 January to 1 February 1992.

Schmink, M. 1992. The socioeconomic matrix of deforestation. Paper presented at the Workshop on Population and Environment, sponsored by Development Alternatives with Women for a New Era, International Social Science Council, and Social Science Research Council, Hacienda Cocoyoc, Morelos, Mexico, 28 January to 1 February.

Schmink, M. and Wood, C. 1987. The political ecology of Amazonia. In *Lands at Risk in the Third World*, ed. P. Little and M. Horowitz. Boulder: Westview Press.

Schmitz, H. 1982. Growth constraints on small-scale manufacturing in developing countries: a critical review. *World Development* 10:6.

Schmitz, H. 1990. Small firms and flexible specialization in developing countries. *Labour and Society* 15:3.

Schulze, R. E. and Lynch, S. D. 1992. Distributions and variability of primary productivity over southern Africa as an index of environmental and agricultural resource determination. Presented at ICID, Fortaleza-Ceará, Brazil, 27 January to 1 February 1992.

Scott, J. C. 1976. *The Moral Economy of the Peasant: Rebellion and Subsistence in Southeast Asia*. New Haven: Yale University Press.

Selvarajan, S. and Sinha, S. K. 1992. Weather variability and food grains production sustainability in India. Presented at ICID, Fortaleza-Ceará, Brazil, 27 January to 1 February 1992.

Sen, A. 1981. *Poverty and Famines: An Essay on Entitlement and Deprivation*. Oxford: Oxford University Press.

Sen, A. 1987. *Hunger and Entitlements*. World Institute for Development Economics Research of the United Nations University.

Serpantié, G. 1992. Should the small areas of 'Bas-Fonds' in the Sudano-Sahelian region be considered within the strategy of agricultural household production. Presented at ICID, Fortaleza-Ceará, Brazil, 27 January to 1 February 1992.

Servain, J., Merle, J. and Morlière, A. 1992. The influence of the tropical Atlantic Ocean on the hydroclimates of the Sahel and the Northeast. Presented at ICID, Fortaleza-Ceará, Brazil, 27 January to 1 February 1992.

Soares, A. M. L., Leite, F. R. B. Leite, Lemos, J. de J. S., *et al.* 1992. Degraded areas susceptible to desertification processes in the state of Ceará–Brazil. Presented at ICID, Fortaleza-Ceará, Brazil, 27 January to 1 February 1992.

Stevens, W. K. 1994. Threat of encroaching deserts may be more myth than fact. *New York Times*, 18 January, pp. C1, C10.

Stone, P. H. 1992. Forecast cloudy: the limits of global warming models. *Technology Review*, MIT, February/March.

Storper, M. 1991. Regional development policy 2: the global industrial economy and local industrial strategies. In *Industrialization, Economic Development and the Regional Question in the Third World: From Import Substitution to Flexible Production*. London: Pion Press.

Swift, J. 1989. Why are rural people vulnerable to famine? *IDS Bulletin*, 20(2), 8–15.

Swindale, L. D. 1992. A research agenda for sustainable agriculture. Presented at ICID, Fortaleza-Ceará, Brazil, 27 January to 1 February 1992.

Tendler, J. 1989. Whatever happened to poverty alleviation? In *Microenterprises in Developing Countries*, ed. J. Levitsky. London: Intermediate Technology Publications.

Tie Sheng, Li, Liu Zhong Lin, Xie Qinyun, Wang Shide and Ma Lanzhong. 1992. The situation, problems, countermeasures of arid, semi-arid regions in Inner Mongolia. Presented at ICID, Fortaleza-Ceará, Brazil, 27 January to 1 February 1992.

Timmer, C. P. 1992. *Agriculture and the State*. Ithaca: Cornell University Press.

Tucker, C. J., Dregne, H. E. and Newcomb, W. W. 1991. Expansion and contraction of the Sahara Desert 1980–1990. *Science* 253. 299–301.

UCAR 1992. *Arid Ecosystems Interactions: Recommendations for Drylands Research in the Global Research Program*. Boulder: University Corporation for Atmospheric Research.

Vallianatos, E. G. 1992. Sustainable development theory. Presented at ICID, Fortaleza-Ceará, Brazil, 27 January to 1 February 1992.

Watts, M. J. 1983. *Silent Violence: Food, Famine and Peasantry in Northern Nigeria*. Berkeley: University of California Press.

Watts, M. J. 1987a. Conjunctures and crisis: food, ecology and population, and the internationalization of capital.' *Journal of Geography* [Nov/Dec].

Watts, M. J. 1987b. Drought, environment and food security: some reflections on peasants, pastoralists and commoditization in dryland West Africa. In *Drought and Hunger in Africa*, ed. M. H. Glantz, pp. 171–211. Cambridge: Cambridge University Press.

Watts, M. J. and Bohle, H. 1993. The space of vulnerability: the causal structure of hunger and famine. *Progress in Human Geography* **17**(1), 43–68.

WCED 1987. *Our Common Future: Report of the World Commission on Environment and Development*. London: Oxford University Press.

Wilhite, D. A. and Glantz, M. H. 1987. Understanding the drought phenomenon: the role of definitions. In *Planning for Drought: Toward a Reduction of Social Vulnerability*, ed. D. A. Wilhite and W. E. Easterling with D. A. Wood, pp. 11–30. Boulder: Westview Press.

Wilhite, D. A. and Easterling, W. E. with Wood, D. A. (eds.) 1987. *Planning for Drought: Toward a Reduction of Social Vulnerability*. Boulder: Westview Press.

Wisner, B. 1976. *Man-Made Famine in Eastern Kenya: The Interrelationship of Environment and Development*. Discussion Paper No. 96. Brighton: Institute of Development Studies at the University of Sussex.

World Bank 1990. *World Development Report 1990*. Washington: World Bank.

WRI 1990. *World Resources 1990–91*. New York: Oxford University Press.

WRI 1991. *World Resources 1991–92*. New York: Oxford University Press.

WRI. 1992. *World Resources 1992–93: A Guide to the Global Environment*. A report by the World Resources Institute in collaboration with UNEP and UNDP. New York: Oxford University Press.

Yair, A. 1992. The ambiguous impact of climate change at a desert fringe: Northern Negev, Israel. Presented at ICID, Fortaleza-Ceará, Brazil, 27 January to 1 February 1992.

II

Climate Variation, Climate Change and Society

2: Climate Change and Variability in Mexico*

KAREN O'BRIEN AND DIANA LIVERMAN

INTRODUCTION

Mexico, located between latitudes 15° and 32°N, spans a range of climates from arid desert to the humid tropics (Fig. 1). There is a significant geographic mismatch between water and human occupancy in Mexico. Seven percent of the land, lying in the extreme southeast of the country, receives 40% of the rainfall. Only 12% of the nation's water is on the central plateau where 60% of the population – including the 18 million people who live in Mexico City – and 51% of the cropland are located. Of Mexico's 195 million hectares (m ha) of land, only 15% is classified as humid or very humid; the remaining 85% is semi-arid, arid, or very arid (Table 1). Thus, in most parts of Mexico, human activity relies on the low, seasonal and variable rainfall that characterizes the arid and semi-arid regions of the world.

The dry climates, together with steep topography, are the main reasons why only 16% of Mexico's land area is considered suitable for crop production (Table 2). In contrast, 38% is considered appropriate for pasture. More than one-third of Mexico's rapidly growing population works in agriculture, a sector whose prosperity is critical to the nation's debt-burdened economy and to national self-sufficiency in food.

One-fifth of Mexico's cropland is irrigated and this area accounts for half of the value of the country's agricultural production, including many export crops. Many irrigation districts rely on small reservoirs and wells which deplete rapidly in dry years. The remaining rainfed cropland supports many subsistence farmers and provides much of the domestic food supply.

Mexico presently generates about one-fifth of its electricity from hydroelectric schemes and has developed less than a quarter of its potential (World Resources Institute 1990). Water supply infrastructure has not been able to keep up

with rapid urban and industrial development. These other sectors compete with the agricultural sector, which consumes more than 80% of total water supplies.

Mexico, and the Mexican people, are extremely vulnerable to droughts and other climatic variations. In this chapter we review some of the literature on climate change and variability in Mexico, and present some of our own work on drought in twentieth-century Mexico and on the possible impacts of global warming. Our focus is on the impacts of past and future droughts on the Mexican agricultural system.

MEXICAN CLIMATE

The mountainous terrain and dissected topography of Mexico result in tremendous climate variability over short spatial distances, with variations corresponding as much to altitude as to latitude. Other permanent controls influencing the climate include land–sea distributions, the influence of offshore ocean currents, and the location of tropical storm tracks. Despite large variations, the climate of Mexico can be divided into three broad categories. The first category consists of the wet, tropical climates that are generally found in southern Mexico and on the Pacific and Gulf coasts, south of latitude 24° N. The second type includes temperate, seasonally moist climates typical of the mountainous areas and central plains. Finally, the third category consists of the dry climates generally found in the northern part of the country, including the Baja California Peninsula and the Pacific coastal plains north of latitude 25° N.

In terms of atmospheric circulation dynamics, the climate of Mexico is strongly influenced by the seasonal shifting of subtropical high-pressure cells and the intertropical convergence zone. The subtropical high-pressure belt creates stable, dry conditions over most of Mexico in winter, when it has shifted to its southernmost position. During this time in the northwestern region of Mexico, westerly winds penetrate

* Several sections of this paper are drawn from Liverman 1992a.

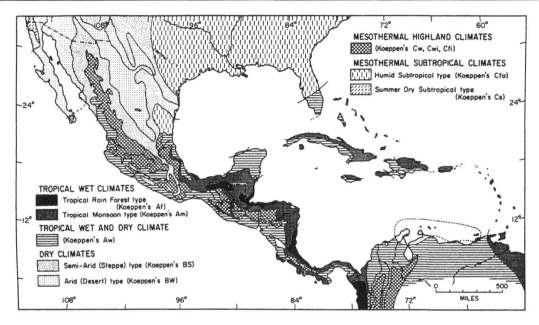

Figure 1: Climatic areas of Mexico. (Source: West and Augelli 1990.)

and bring some winter rainfall. In the summer, as the subtropical high-pressure belt shifts northward, the northwest region experiences hot, dry conditions. Convection and the moist, steady trade winds bring heavier rainfall to eastern and central Mexico and the Yucatan peninsula. Precipitation reaches a maximum in the summer, as the intertropical convergence zone moves north. Heavy rainfall is also associated with hurricanes, which tend to form in late summer and early fall. In winter, cold polar air, or *nortes*, can surge southward into Mexico, bringing rain, on the Gulf Coast, and frost in the highlands. In midsummer, high pressure sometimes develops over the central plateau, disrupting the easterly flow and causing a period of drought known as the *canicula*. The mountainous regions along the spine of the country and the central plateaus tend to be cooler, with higher precipitation on the windward side (Mosiño and Garcia 1973; Mosiño 1975; Metcalfe 1987).

Meteorological records are available for some parts of Mexico since the nineteenth century, and 30-year normals are available for over 300 climate stations (Servicio Meteorologico 1976). The lowland regions of Mexico experience annual average temperatures of around 20 °C. The highlands and the north have larger daily and annual temperature ranges, with annual averages decreasing with increasing altitude. Annual rainfall totals vary widely, from below 100 millimeters (mm) a year in the desert northwest to over 2000 mm a year in the south and east.

With the exception of the southern and Gulf Coast regions, much of Mexico experiences arid and semi-arid conditions, characterized by sparse and highly variable rainfall. Coefficients of variation of annual rainfall are higher

than 30% over much of the country. The vegetation of the arid and semi-arid zones of Mexico is characteristic of desert and grassland, with great diversity relative to other arid regions of the world (Fig. 2).

HISTORY OF DROUGHT IN MEXICO

Paleoclimatic and instrumental records indicate that there have been significant shifts in the climate of Mexico throughout history, including long periods of drought (Metcalfe 1987). Fossil evidence suggests that the arid zones of northern Mexico expanded and contracted during past glacial and interglacial periods, occupying smaller areas in the past than at present (Ezcurra and Montaña 1990). Marine and pollen records support this, showing that relatively arid periods occurred from 2000 to 1800 BP (Before Present, defined 1950), and from 1000 BP to the present. The increasing appearance of maize pollen between 6500 and 4500 BP, following a transition from moist to dry conditions (Brown 1985), has led some scientists to believe that the domestication of plants for agricultural use was tied to climate change.

Together with changes in social organization, climate changes have been implicated in both the rise and the collapse of powerful pre-Hispanic civilizations, such as those in the Yucatan and in the valley of Mexico. Past and present Mexican society has adapted to the constraints of the arid and semi-arid regions. Indigenous groups, such as the Yaqui and Papago of the Sonoran desert, grew crops adapted to the desert environment, and used water-harvesting techniques (Dunbier 1968). In other parts of Mexico, many different

Table 1. *Classification of Mexican land area by general climate type*

Climate	Area (ha)	%
Very humid	4901	2
Humid	26213	13
Semi-arid	74072	38
Arid	54226	28
Very arid	36988	19

Source: Reyes Castaneda (1981).

Table 2. *Land use potential in Mexico*

Use	Area (ha)	%
Arable	30350	15.6
Pasture	75850	37.9
Forest	44170	22.5
Uncultivated	10610	5.5
Non-agricultural	36410	18.5

Source: Reyes Castaneda (1981).

agricultural technologies – such as the *chinampa* system and terracing – have been developed to conserve water and reduce drought risks (Wilken 1987).

With Spanish colonization there was a tremendous expansion of irrigation in Mexico, which extended the range and reliability of crop production while increasing competition for limited water supplies in some regions. Historians have blamed climate for famine and social unrest during the colonial period. Between 1521 and 1821, 88 droughts were recorded, with the worst in 1749, 1771 and 1785. Florescano (1969) has linked variations in the price of maize in the sixteenth and seventeenth centuries to droughts and other climatic events. Florescano claims that the majority of price rises were preceded by a severe drought, but he also notes the role of speculation and economic arrangements in triggering price rises and associated famines. He suggests that the economic and land-tenure relations imposed by the Spanish crown and church created a tremendous vulnerability to drought among the poorer and indigenous *campesino* populations. The colonial political economy allowed the larger land-holders and merchants to manipulate the price of staples in drought years to the disadvantage of poor consumers and small producers.

The Spanish introduced cattle into arid northern Mexico (Fig. 3), beginning the process of desertification from overgrazing which continues to the present day. Widespread deforestation and desiccation were also associated with the development of colonial mining settlements in regions such as Durango and Zacatecas (West and Augelli 1990).

DROUGHT IN TWENTIETH-CENTURY MEXICO

Since independence and the Mexican Revolution, a steady and unvarying expansion of Mexican agricultural production has been necessary to meet the demands of a rapidly growing population and agricultural export market. In the drive to modernize and expand production, the Mexican

agricultural system has incorporated techniques which may reduce vulnerability to climate variations, such as irrigation, fetilizers, pesticides and improved seed varieties.

However, some believe that the large regions of high-yielding monocultures established through Green Revolution technologies are much more vulnerable to pests and diseases than traditional mixed cropping (Pearse 1980). It is also argued that the new techniques have replaced some traditional hazard prevention strategies, such as mixed cropping and microclimate modification (Wilken 1987), and have enabled agriculture to expand into areas of high hazard risk such as deserts, mountains, coastal regions, and the disease-susceptible humid tropics.

In the debate concerning the redistribution of land that followed the revolution, the performance of the *ejido* sector, consisting of communally held plots, has been a particular point of contention. More than 50% of Mexico's cultivated land is held in *ejidos*. Although some authors suggest that the *ejidos* are inefficient and unproductive (Wellhausen 1976), others claim that, in terms of input use, they are relatively efficient and produce yields equal to the private sector (Dovring 1970; Nguyen 1979). The problems associated with the *ejido* sector have been attributed to their lack of political power, difficulties in obtaining access to credit and inputs, poor management, and low-quality land resource endowments (Hewitt de Alcantara 1976; Coll-Hurtado 1982).

Mexican agricultural practices vary from small, traditional farms to large, high-input corporate farms. Associated with these, and inherited from the colonial land-tenure systems, are differential vulnerabilities to drought. Florescano (1980) suggests that, as in earlier times, the most disastrous effects of drought are concentrated in the rainfed agriculture practiced by the poorest *ejidatarios* and *campesinos*, who lack credit, irrigation, fertilizers and improved seeds.

Our research has shown widespread and severe vulnerability to drought and other hazards in the twentieth century, such as floods and frosts, which are associated with climate variability (Liverman 1992*a*). The natural hazard losses in

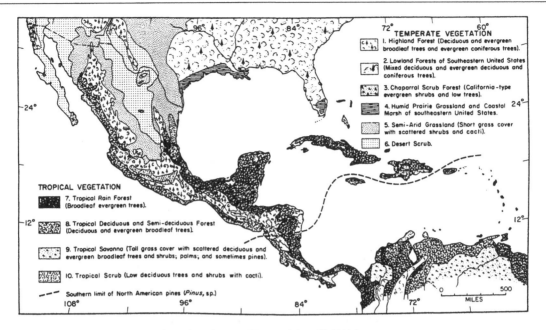

Figure 2: Natural vegetation of Mexico. (Source: West and Augelli 1990.)

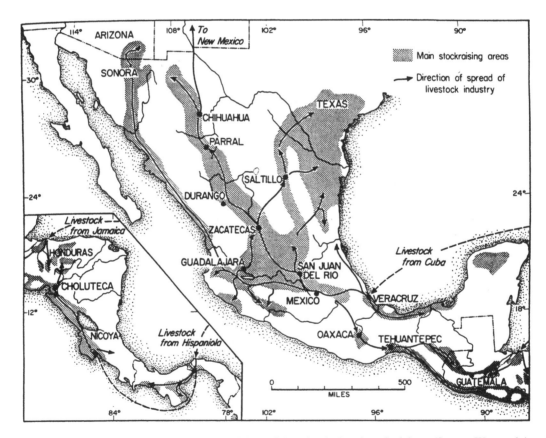

Figure 3: The spread of livestock economy in Mexico and Central America during the colonial era. (Source: West and Augelli 1990.)

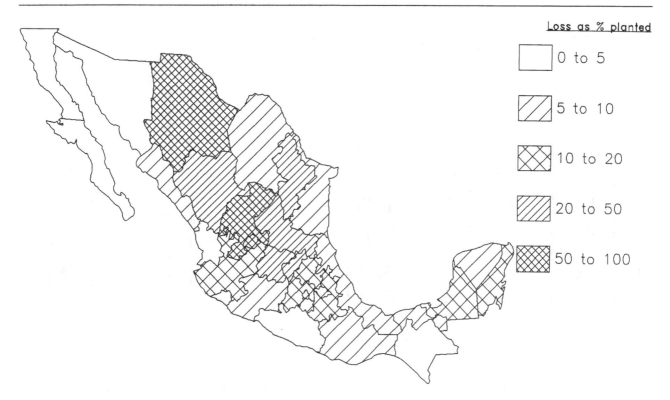

Figure 4: Drought loss (%) in Mexico 1970.

agriculture from 1930 to 1980 at the national level in Mexico have been analyzed using data from the agricultural census. On average, more than 90% of hazard losses in Mexican agriculture are from drought. Furthermore, the area of total hazard losses increased in Mexico from 1930 to 1970, and the total drought losses increased from 1950 to 1960 to 1970. This partly reflects an increase in the total crop area, from 7.8 m ha in 1930 to 11.8 m ha in 1960. Relative (percentage) losses also increased, indicating that some of the land expansion may have included more hazard-vulnerable land. However, both absolute and percentage hazard losses increased dramatically in 1970 when land area dropped to 10.2 m ha. The meteorological record does not indicate an increasing severity of weather events in the census years, or from 1960 to 1970. The increase may, therefore, support a hypothesis that hazard losses have been increasing irrespective of weather severity, because of increases in social vulnerability to natural disasters.

High drought losses tend to occur in the northern region of Mexico, where precipitation is low and highly variable. Fig. 4 shows that parts of northern Mexico lost more than 35% of the area planted to drought in the summer of 1970. In many regions of local rainfed agriculture, drought destroys more than 50% of the crops that are planted, and farmers risk harvest failure 1 in every 3 years.

One of the major buffers against drought in arid and semi-arid regions is irrigation. However, if water supplies are limited, or climate changes, reliance on irrigation can be risky. Is there any evidence that irrigation buffers agriculture against drought at the national level? States with high levels of irrigation, such as Sonora and Sinaloa, do have much lower drought losses than many of the states with smaller proportions of irrigated land. However, in 1970, Aguascalientes and San Luis Potosi, with about 20% of their land irrigated, reported drought losses of 70% and 50%, respectively, indicating that irrigation may not always buffer Mexican states against drought.

As noted earlier, some people suggest that the use of improved seeds, such as those produced at CIMMYT (Centro Internacional de Mejoramiento de Maiz y Trigo), may be associated with changes in natural hazard vulnerability. We have found a correlation between expenditure on improved seeds and drought losses to be negative but weak (-0.18 in 1970), suggesting that higher expenditures on improved seeds may be associated with lower drought losses.

As noted earlier, the relative economic vulnerability of the *ejidos* has become a matter of considerable academic and political interest. Fig. 5 shows the average losses, at the state level, for the two main land-tenure sectors in Mexico. In every census year the average losses in the private sector are less than those in the *ejido* sector. In 1950, losses were double on *ejido* lands. Differences between private and *ejido* losses are significant for each hazard, with *ejidos* generally reporting higher drought, flood, and frost losses.

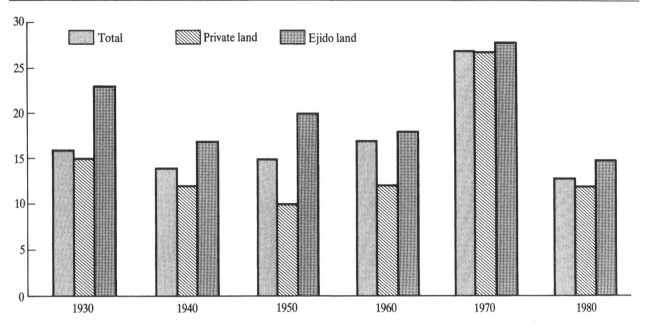

Figure 5: Average losses for the two main land-tenure sectors in Mexico.

These results seem to confirm the ideas of Florescano about the relative vulnerability of the *ejido* sector to drought and to support the findings of a previous study of drought losses at the *municipio* level in Sonora and Puebla in 1970 (Liverman 1990). Explanations of this increased vulnerability are that more biophysically marginal land was given to *ejidos* in the land reform, and that *ejidos* are socially vulnerable because they cannot get access to irrigation, credit, improved seeds or other resources.

GLOBAL WARMING

It is clear that Mexico is already severely affected by climatic variability, and is particularly vulnerable to drought conditions. Many scientists believe that an increase in carbon dioxide (atmospheric CO_2) is likely to result in warmer global temperatures and altered precipitation patterns – scenarios which could have serious implications for the arid and semi-arid climates of Mexico. The majority of global warming experiments with climate models have simulated current climate using a CO_2 concentration of about 300 parts per million (ppm) and have compared the results with a simulation based on a doubling of CO_2 levels to about 600 ppm. Scientists estimate that this doubling of CO_2, which is likely to occur in the first half of the next century, may raise the global mean temperature by 1.3–5.1 °C (IPCC 1990). The magnitude of temperature changes will vary across regions of the globe and will be accompanied by changes in rainfall and other climatic variables. The poles, for example, are expected

to warm more than the equator, and precipitation will increase in some places and decrease in others.

What do the models project for the climate of Mexico? We obtained the $1 \times CO_2$ and $2 \times CO_2$ results for the grid points corresponding to the Mexican landmass from five major general circulation models (GCMs) (Liverman and O'Brien 1991). The resolution of the models is represented on grids superimposed on a map of Mexico in Fig. 6. The models we refer to are:

(1) The *Geophysical Fluid Dynamics Laboratory* (GFDL) model developed at Princeton and described by Manabe and Wetherald (1987). The results used in the Mexico study are from the 'Q' run which includes variable clouds and has a 7.5° × 4.44° geographic grid.
(2) The *Goddard Institute for Space Studies* (GISS) model described by Hansen *et al.* (1988). We use the 'Model 1' equilibrium run result for a 10° × 7.83° grid.
(3) The *National Center for Atmospheric Research* (NCAR) model described by Washington and Meehl (1986). We use the mixed layer ocean version and seasonal rather than monthly results on a 7.5° by 4.44° grid.
(4) The *Oregon State University* (OSU) model discussed by Schlesinger and Zhao (1989) with a mixed layer ocean and a 5° × 4° grid.
(5) The *United Kingdom Meteorological Office* (UKMO) model described by Wilson and Mitchell (1987) with a grid 7.5° × 5°.

These models vary in their assumptions about cloud cover, ice feedbacks and the representation of oceans, and in their characterization of topography, geography and soil depth,

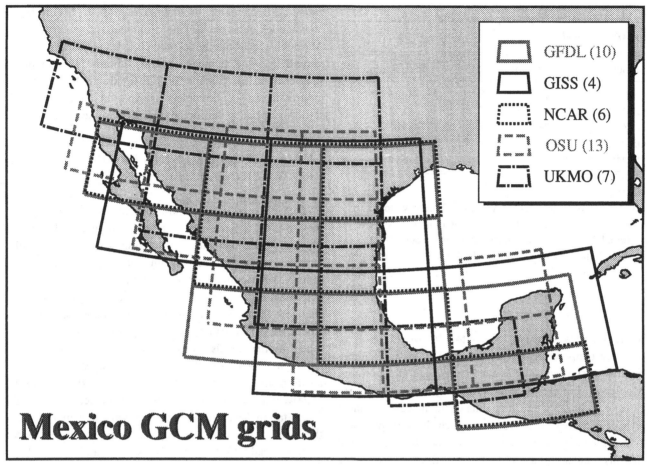

Mexico GCM grids

DEASY GEOGRAPHICS LAB

Figure 6: Mexico GCM grids. For details of models see text. The numbers in parentheses are the number of grid squares that cover Mexico in each model. (Source: Deasy Geographics Lab.)

among other factors. In spite of these differences, there are many similarities in the structure and parameterizations among the models (Bach *et al.* 1985; Schlesinger and Mitchell 1987).

GLOBAL WARMING PROJECTIONS AND MEXICAN CLIMATE

The models project annual average temperature increases ranging from 2.38 °C to 5.44 °C (Table 3). The version of the UKMO model used in this study produces greater temperature increases than the other models due to differences in cloud feedback mechanisms, which result in higher net radiation and lower soil moisture (Wilson and Mitchell 1987). These temperature increases are much greater than the variations estimated to have occurred in Mexico over the last several thousand years. They are similar in magnitude to the warming forecast for much of the mid-latitudes, including

the United States and Europe. The temperature changes projected by the models are much more consistent than those for precipitation. The model results for precipitation vary widely, from a 23% decline in Mexican annual average precipitation projected by the NCAR model to a 3% increase projected by the GISS model.

Fig. 7 shows current and projected changes in climate at a number of semi-arid and arid meteorological stations in Mexico. Temperatures under global warming have been estimated by taking the temperature change projected by the models for the grid square in which the station is located, and adding the increment to the observed 30-year mean. Precipitation estimates are based on the percentage change from the $1 \times CO_2$ to $2 \times CO_2$ simulations, multiplied by the observed monthly precipitation.

For all of the stations, all five of the models project increases in temperature for every month of the year, resulting in more months of extremely high summer temperatures. The changes are greater for locations in northern Mexico,

Table 3. *Annual average temperature changes predicted by five GCM models*

Model	Temperature (°C)	Precipitation (%)
GFDL	3.11	− 1.75
GISS	3.92	2.74
NCAR	2.38	− 23.01
OSU	3.15	− 1.05
UKMO	5.44	− 0.09

Notes:
GFDL, Geophysical Fluid Dynamics Laboratory model;
GISS, Goddard Institute for Space Studies model; NCAR,
National Center for Atmospheric Research model; OSU,
Oregon State University model; UKMO, UK Meteorological
Office model.

where monthly increases of up to 7 °C would result in monthly temperatures greater than 30 °C during summer months at stations such as La Paz, Ciudad Obregon and Monterrey. For example, mean monthly summer temperatures of about 25 °C in Chihuahua could increase to almost 30 °C (Fig. 8). In Ciudad Obregon, which already experiences summer temperatures near 30 °C, temperatures in several months could approach 35 °C (Fig. 9). These increases in the already hot summers of northern Mexico are likely to bring great stress to humans, flora and fauna, particularly if these monthly means imply even higher daily temperature extremes. On the other hand, warmer temperatures may extend the growing season at higher elevations, such as Chihuahua, which sometimes experiences cold temperatures and frosts during winter months. Temperature increases are not quite as high in central Mexico as in the north. Mexico City's current winter temperatures of around 12 °C could increase to as high as 18 °C. In the highland agricultural regions around Mexico City, such as Puebla, warmer temperatures could be beneficial by reducing the risk of frost.

The range of projected precipitation changes is much greater than that for temperature, with increases shown for some models and some months and decreases for others. In the arid northern cities of La Paz, Tijuana, Ciudad Obregon and Chihuahua, several models project higher summer rainfall, but others suggest an overall decline in annual rainfall with global warming. For example, in Ciudad Obregon, which is located in one of the most important irrigated agricultural regions of Mexico, the GISS and UKMO models indicate that rainfall may increase in July and August, whereas other models show lower rainfall throughout the year (Fig. 10). Scenarios for Chihuahua's climate follow similar patterns, but with significant drops in OSU

and GFDL estimates for summer rainfall. In Monterrey, most models indicate a decline in summer and total rainfall. Although at least one model predicts a dramatic increases in rainfall in some summer and fall months in the moist southern climates of Jalapa, San Cristobal and Merida, other models suggest significant decreases in total rainfall at many stations, especially during the summer rainy season.

LIMITATIONS OF USING MODEL PROJECTIONS

There are several criticisms and limitations associated with the use of GCMs in assessing the regional impacts of global warming. Although most model results show an overall increase in temperatures, they differ in their regional projections of the magnitude of the warming, and often show little agreement on the direction and magnitude of precipitation changes. Many key processes and components, such as cloudiness and ocean circulation, are inadequately represented by the models. The coarse scale of the model grids makes it difficult to allocate climate changes to specific locations and tends to neglect some of the important sub-grid scale weather patterns that can be important determinants of precipitation. Several studies have shown that in many cases the models are unable to replicate current climate and climate variability (Grotch 1988; Kellogg and Zhao 1988; IPCC 1990).

ABILITY OF THE MODELS TO REPRODUCE OBSERVED MEXICAN CLIMATE

The models perform rather poorly in simulating the current climate of Mexico. Comparisons between model results for 'current' or '$1 \times CO_2$' levels of carbon dioxide (approximately 300 ppm) and observed climates are complicated by differences in, and the large size of, the spatial grids of the models (Fig. 7), as well as the uneven distribution of meteorological stations. One of our methods of comparison is based on individual meteorological stations, comparing the most recent climatic normals (e.g. 1951–1980 average temperature and total precipitation) with the average temperature and precipitation produced by the $1 \times CO_2$ simulation of the models for the grid square in which the station is located. Figs. 11 and 12 show the results of such a comparison for Chihuahua, located in the arid north of Mexico. In warm, arid Chihuahua, the models estimate annual average temperatures ranging from − 2 °C to + 6 °C of the observed, with the UKMO model coming closest to observed conditions. The UKMO model also comes the closest to replicating observed total annual rainfall at Chihuahua, in contrast

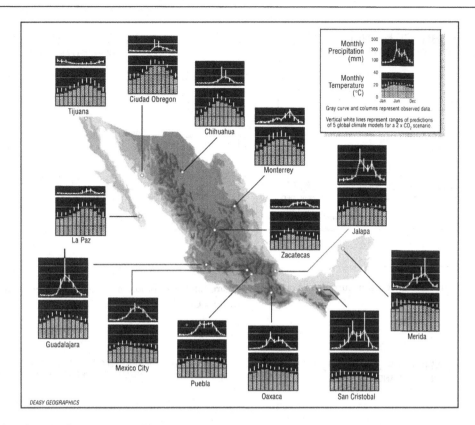

Figure 7: Current and projected climate changes in Mexico. (Source: Deasy Geographics Lab.)

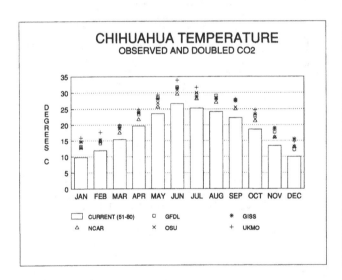

Figure 8: Mean monthly temperatures in Chihuahua: current and projected with $2 \times CO_2$.

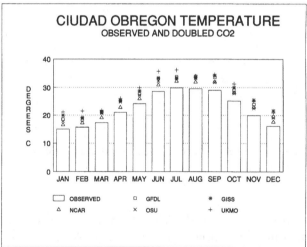

Figure 9: Mean monthly temperatures in Ciudad Obregon: current and projected with $2 \times CO_2$.

to the GFDL and NCAR models, which seriously over-estimate it (Fig. 12).

The inability of the models to reproduce the seasonality of rainfall in northern Mexico represents a serious problem. All models, except GFDL, show a winter maximum of rainfall in Chihuahua, whereas observed maximum rainfall occurs in the summer. Possible explanations for this discrepancy

include: (1) the models are generating the westerlies (which bring winter rainfall only in the far northwest of Mexico) much further south than actually occurs; or (2) the lack of topography in the models is obscuring the observed rainsha-dow effect of the Sierra mountains in winter.

Individual meteorological station conditions may not be representative of the regional climates which correspond to

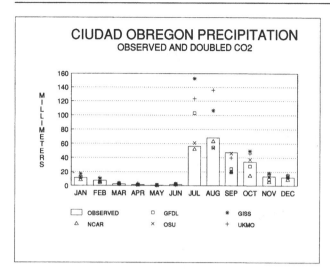

Figure 10: Precipitation in Ciudad Obregon: current and projected with $2 \times CO_2$.

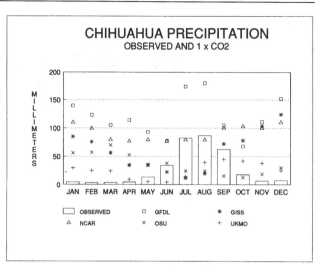

Figure 12: Precipitation in Chihuahua: observed and projected with $1 \times CO_2$.

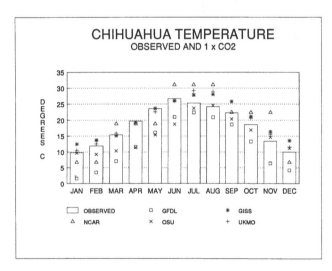

Figure 11: Temperature in Chihuahua: observed and projected with $1 \times CO_2$.

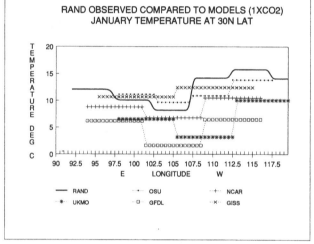

the scale of the GCM grids. The heterogeneity of the Mexican environment means that meteorological observations within a model grid square may vary widely. To investigate this, we have used regional averages of observed climates developed by the RAND Corporation, in which temperature and precipitation values are averaged for meteorological stations within a $5° \times 4°$ geographic grid. The RAND dataset is that being used by the United States Environmental Protection Agency (USEPA) to compare GCM control runs with observed data (USEPA 1991).

Fig. 13 shows model results compared with RAND climatology for January temperatures and annual precipitation, across an east–west transect at latitude 30° N. Observed conditions in January show a general increase in tempera-

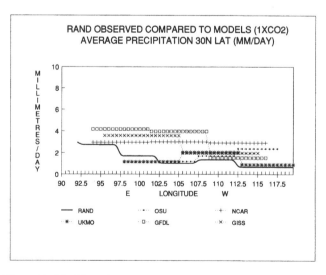

Figure 13: Model results compared with RAND climatology for January temperature and precipitation at latitude 30° N.

tures from east to west, but with a decline around 105°
longitude which represents the cooling influence of higher
elevation stations. The NCAR and OSU models follow this
climatology most closely; the UKMO and GFDL models
produce temperatures that are too low, even in the highland
region. The RAND annual rainfall declines from east to west
across northern Mexico, from the more humid Gulf Coast to
the deserts along the Pacific. There is a slight increase in
observed rainfall as air starts to rise over the highlands at
longitude 110° W. The GISS and GFDL models show
rainfall decreasing from east to west, but the OSU model
produces an opposite trend. The UKMO model seems to
replicate best the orographic rainfall increase over north-
central Mexico.

These results demonstrate some of the inadequacies in the
ability of climate models to reproduce observed climates, and
can lead to distrust in the changes projected by the models,
particularly those for precipitation. Although the $1 \times CO_2$
simulations do not always correspond with the regionally
averaged observed climates, the agreement is generally closer
than with the individual station values. It may, therefore, be
appropriate to compare GCM results for the current climate
with such spatial averages of observed conditions. Overall,
the UKMO model seems to provide the best representation
of contemporary Mexican climate.

Table 4. *Changes in potential evaporation and water availability*

	Observed	Model				
		GFDL	GISS	NCAR	OSU	UKMO
Ciudad Obregon						
PE	2437 mm	19%	20%	8%	13%	27%
P – PE	– 2181 mm	– 21%	– 16%	– 12%	– 15%	– 23%
Merida						
PE	2483 mm	9%	9%	6%	8%	12%
P – PE	– 1496 mm	– 3%	– 5%	– 7%	– 2%	24%
Mexico City						
PE	1782 mm	7%	10%	6%	8%	10%
P – PE	– 966 mm	– 23%	– 23%	– 11%	– 14%	– 15%
Monterrey						
PE	2106 mm	18%	19%	13%	17%	31%
P – PE	– 1515 mm	– 29%	– 28%	– 33%	– 23%	– 47%
Oaxaca						
PE	2123 mm	5%	12%	3%	8%	16%
P – PE	– 1516 mm	– 2%	– 19%	– 6%	0	– 22%
Puebla						
PE	1528 mm	14%	12%	7%	10%	16%
P – PE	– 694 mm	– 45%	– 33%	– 23%	– 18%	– 30%

Notes:
PE, potential evaporation (Penman); P – PE, Precipitation minus
potential evaporation; models as in Table 3.

SOIL MOISTURE AND WATER AVAILABILITY

Assessing changes in moisture availability, particularly in
locations where both temperature and rainfall are projected
to increase, requires the analysis of evaporation. We need to
know how any increase in rainfall or growing season length is
likely to interact with the increased evaporation associated
with higher temperatures. Potential evaporation is the
amount of water that would be evaporated into the atmosphere from a completely wet surface and depends on factors
such as temperature, solar radiation, humidity and wind.
Most of the models calculate evaporation and soil moisture
internally, but the surface energy and water budgets of the
models tend to be rather simple. For example, all but the
GISS model represent soil water using a 15 centimeters deep
'bucket,' which when full produces runoff. Moisture is lost
through evaporation, at slower rates as the bucket empties
(Kellogg and Zhao 1988). A problem arises in dry regions
where the $1 \times CO_2$ simulation of soil moisture is low. Even if
potential evaporation in the model rises under the higher
temperatures of $2 \times CO_2$ there is little or no water to evaporate in the bucket and, as a result, the estimated change in soil
moisture will be very small. In reality, deeper soils can often

contain more water, and soil moisture could decline significantly as temperatures rise.

The water balance results of the GISS and OSU models
project that potential evaporation will increase in all months
because higher air temperatures increase the atmospheric
demand for moisture in air that is relatively dry. The GISS
model internally estimates a 15% increase in potential evaporation for the Chihuahua grid square. The OSU model, on
the other hand, projects a 12% decrease in potential
evaporation.

Because of the limitations of modelling soil moisture in the
GCMs, we have used model projections for precipitation,
temperature and solar radiation changes to estimate evaporation and moisture availability exogenously using the
Thornthwaite and Penman methods. The Thornthwaite
method uses temperature, whereas the Penman procedure
also includes radiation, humidity and wind speed. Table 4
shows changes in Penman potential evaporation projected
by the different models for a selection of Mexican locations.
For these particular analyses, no change in wind, relative
humidity or cloudiness was assumed.

Potential evaporation is projected to increase at all the
stations because higher air temperatures increase the atmospheric demand for moisture in air that is relatively dry. For
example, at Ciudad Obregon, in the arid northwest, current
potential evaporation is estimated at almost 2500 mm per

annum. Depending on the model, this is projected to increase by 8–27%. Monterrey also has large projected increases. These stations in northern Mexico are adjacent to important irrigation districts where increased potential evaporation would increase losses from surface water supplies and irrigated crops. Similar increases in potential evaporation are possible in central Mexico, where demand in locations such as Mexico City, Puebla and Oaxaca could increase by 5–15%.

The relationship between increased potential evaporation and changes in precipitation can be simply approximated over a year by subtracting total annual potential evaporation from annual precipitation to produce an estimate of moisture surplus or deficit. Under current climate, such a calculation for Ciudad Obregon results in a moisture deficit of 2181 mm. Although the GISS and UKMO models project increases in precipitation in northwest Mexico, these are counteracted by the increased potential evaporation associated with higher temperatures, especially in the UKMO case. The moisture deficit increases by 16% with the GISS model, and by 23% for the UKMO model. Similar decreases in water availability are indicated for Monterrey, where the moisture deficit increases from 23% to 47%.

These results suggest that dry regions of northern Mexico may become even drier with global warming, at least as averaged over the year. The consequences for rainfed agriculture, urban and industrial water supplies, and natural ecosystems would be serious. In northwest Mexico, the economy and livelihoods depend on adequate supplies of water to cities and irrigated agriculture. Northern Mexico is experiencing dramatic industrialization and urbanization with ever-increasing demands for water. Three-quarters of the water needs in this region depend on limited supplies from reservoirs and rivers. In 1987, a severe drought year in much of Mexico, reservoir levels in the northwest dropped on average to only 30% of their capacity with a corresponding drop in production of wheat and other important crops. The northwest is also a region where international conflicts over water from the Colorado River could be exacerbated by increases in potential evaporation (Gleick 1988).

Rainfed agriculture and economic development are already limited by annual water deficits in central Mexico. Harvests often fail through a combination of high temperatures and lower than expected rainfall. The Puebla region, where an 18–45% decline in water might occur, is an important region of agricultural employment and supplier of food to Mexico City. Mexico City has a growing demand for water, which is currently met by bringing in water from the surrounding area. If water supplies shrink by 11–23% in the basin of Mexico, there will probably be serious implications for health and economic growth in Mexico City.

In Oaxaca, where many peasant farmers have to deal with recurring droughts, some optimism is offered by the GFDL model, where the moisture deficit is estimated to be reduced by 2% compared with current conditions. In this case, the projected increase in precipitation in the Oaxaca grid square seems to compensate for the higher potential evaporation produced by increased temperatures. However, the GISS and UKMO models, which project precipitation decreases in Oaxaca, produce 19% and 22% increases in water deficits, respectively.

The implications of possible errors in the model projections for precipitation and solar radiation have been investigated through a sensitivity analysis of the Penman calculations. For example, in the cases of Puebla and Ciudad Obregon we find that it would require significant decreases in wind speeds or solar radiation to prevent an increase in potential evaporation with higher temperatures. In Puebla, the effects of global warming under the GFDL scenario would be offset by a 25% decrease in solar radiation, a 50% decrease in wind speed, or a 25% increase in relative humidity. Moisture deficits could only be reduced by much larger increases in rainfall than currently projected by the models. For example, to increase water availability under the GFDL scenario, precipitation would need to increase by 25%. Only if the climate models are greatly overestimating temperature increases, or if they are very wrong regarding rainfall and solar radiation, does the future seem optimistic for water availability if global warming occurs in Mexico.

IMPLICATIONS FOR MEXICAN AGRICULTURE

We have also assessed the possible impacts of global warming on Mexican crop production using the CERES-MAIZE crop yield model. The CERES model uses daily weather data to estimate the effects of climate on yields of rainfed or irrigated maize. The model is physiologically based, simulating the influence of solar radiation, degree-days, nutrients and water availability on the development of maize through major phenological stages (Jones and Kiniry 1986; Adams et al. 1990).

It has been suggested that many crops will benefit from higher level of atmospheric CO_2 and that higher yields may result. The CERES model can also include the direct physiological effects of elevated CO_2 levels on crop growth and water use efficiency. However, the model does not account for changes in pests and diseases and possible yield losses, and may therefore overestimate yields.

The results from one semi-arid agricultural region in the highlands of Mexico will be discussed here. This site, Tlaltizapan, is located to the south of Mexico City in the state of Morelos, at latitude 18°41′ N, longitude 99°08′ W, and at an

Table 5. *Maize yields and climate change at Tlaltizapan*

Scenario	Rainfed yield (tons/ha)		Irrigated yield (tons/ha)	
	330 ppm CO_2	555 ppm CO_2	330 ppm CO_2	555 ppm CO_2
Baseline	4.02	3.81	3.74	2.79
GISS	3.11	3.07	3.10	2.84
GFDL	2.74	3.20	3.38	2.98
UKMO	2.34	1.56	2.17	2.02

Notes:
Models as in Table 3.

altitude of 940 meters. Average annual precipitation in Tlaltizapan is 103 mm, and average temperature is 24 °C. The site has clay-silt soils of average depth. Because soils in Mexico tend to be nutrient-poor, as well as extremely dry at the beginning of the growing season, we modified initial soil conditions to represent low soil nitrogen and soil water. We also reduced the organic content of the soil by reducing the amount of crop residue in the initial conditions. We used a planting density of three plants per square meter, and chose a typical planting date of 20 May. In cases where we applied irrigation we assumed a 0.5 meter irrigation depth at 50% efficiency.

Our baseline results for 16 years at Tlaltizapan produced an average rainfed yield of 4022 kg/ha, and an average irrigated yield of 3740 kg/ha. Irrigated yields are lower because the added water flushes some of the limited nutrients out of the soil and the higher nitrogen stress outweighs the lower drought stress (Table 5). Average observed farm yields in the region of Morelos are about 1500 kg/ha.

Climate-change scenarios were generated using results of the GFDL, GISS and UKMO climate models for changes in temperature, precipitation and solar radiation under doubled CO_2 conditions. Rainfed and irrigated yields are then simulated under these changed climates with higher levels of CO_2 (555 ppm), and compared with the yields estimated for the years of observed weather. The results of these experiments, which initially assumed no alterations in planting dates or other input conditions as the climate changes, are shown in Table 5.

At the Tlaltizapan site, average yields of rainfed maize for the period 1974–89 are estimated at 4022 kg/ha. All rainfed climate-change scenarios (at 555 ppm) result in reductions in maize yields, ranging from a 20% drop for the GFDL model to a 61% drop for the UKMO model. These low yields occur because higher temperatures accelerate the phenology and the crop matures faster with less time for grain filling. There is also high drought stress early in the growing season. In the

Table 6. *Maize yields and climate change at Tlaltizapan: GISS transient scenario*

Scenario	Rainfed yield (tons/ha)	Irrigated yield (tons/ha)
Baseline	4.02	3.74
GISS 2010, 440 ppm CO_2	3.67	3.42
GISS 2030, 550 ppm CO_2	3.32	2.99
GISS 2050, 660 ppm CO_2	2.95	2.22

Note: GISS, Goddard Institute for Space Studies model.

UKMO scenario, yields drop dramatically because an increase in rainfall leaches nutrients out of the soil.

Irrigated yields for the period 1974–89 at Tlaltizapan average 3740 kg/ha, with 129 mm of irrigation water applied during the growing season. Irrigated yields also fall under changed climates by 20% for GFDL, 24% for GISS and 65% for UKMO. Irrigation water needs do not necessarily increase with a warmer climate because of shorter growing seasons. For the GISS and UKMO models, irrigation decreases to 126 mm and 101 mm, respectively; for GFDL, irrigation increases slightly to 134 mm.

Yields of rainfed and irrigated maize have also been simulated at Tlaltizapan for the transient scenario of the GISS model. This provides 'snapshots' of climate and CO_2 levels at three time periods: 2010, 2030 and 2050. The results of these simulations are shown in Table 6. Rainfed and irrigated yields decrease consistently as the climate changes and CO_2 levels increase at Tlaltizapan.

Sensitivity analyses and adaptation experiments indicate the importance of some of the current uncertainties in estimating the impacts of future climatic changes. The results reported above assume no change in the average date of planting or in the varieties. Clearly, as climate changes, farmers and governments will try to adapt to the changes. We

Table 7. *Maize yields and climate change at Tlaltizapan: no nutrient limitations*

Scenario	Rainfed yield (tons/ha)		Irrigated yield (tonnes/ha)	
	330 ppm CO_2	555 ppm CO_2	330 ppm CO_2	555 ppm CO_2
Baseline	4.49	4.60	4.72	4.72
GISS	3.49	3.77	3.85	3.85
GFDL	3.07	3.47	3.97	3.97
UKMO	3.92	3.93	3.93	3.93

Notes:
Models as in Table 3.

have examined whether earlier and later sowing dates may offset the negative impacts of a warmer climate. We have also tried changing the crop variety to see whether different varieties may adapt better to global warming. It is also possible to design hypothetical genetic coefficients within the CERES model that maintain high yields under global warming, but it is not clear that these coefficients are biologically possible. The sensitivity of the model results to these factors will indicate possibilities for adjustment to global warming through altering the dates of planting and the genetic type or characteristics of the seed.

The amount of fertilizer and irrigation are critically important to the maize yields predicted by the CERES models. The experiments above all assume relatively low fertilizer use because many Mexican producers can afford to use only 50–100 kg/ha of nitrogen fertilizer at planting. If more fertilizer becomes available to more farmers as the climate changes, then some of the yield reductions might be offset. Only with adequate water and fertilization can the direct effects of CO_2 be of significant advantage to plants such as maize.

When we run observed and model scenarios with no water or nutrient limitations we find that the baseline scenario produces 4720 kg/ha in Tlaltizapan compared with 3740 kg/ha for the nutrient-limited, irrigated, baseline case (Table 7). The full nutrient scenario of GISS (at 555 ppm) yields 3850 kg/ha, GFDL 3970 kg/ha and the UKMO model 3930 kg/ha. The lack of nutrient limitations is particularly important for the UKMO case, where higher rainfall in the base scenario heavily leaches nutrients from the soil.

However, given the environmental and economic constraints and trends in agricultural inputs in Mexico, unlimited water and nutrients are extremely unlikely. The experiment illustrates, however, the sensitivity of maize yields to assumptions about future resource availability and that yield declines caused by global warming might be mitigated by the increased availability of water and fertilizer.

Table 8. *Rainfed maize yield and climate change: Tlaltizapan planting date and variety adaptations*

Scenario	Tlaltizapan (tonnes/ha)
GISS 555 ppm	3.07
Planting date + 10	1.05
Planting date − 10	2.56
Variety Pioneer	3.53
Variety Suwan	3.53

The dates of sowing might also be an adaptation to global warming. Table 8 shows the effects on rainfed yields of hastening or delaying planting by 10 days and reveals that changing the planting date does not seem to help adapt to climate change. Changing the variety or genetic characteristics of the plant might also assist in adapting to global warming. Planting the unmodified Suwan or a Pioneer variety produces slightly higher yields with the GISS scenario than when using the modified variety at Tlaltizapan.

In this study we found that, in spite of differences between the climate model results, the general direction of changes in Mexican maize rainfed and irrigated yields with global warming is a decrease, whatever the model. Sensitivity analysis indicates that decreases in crop yields will be severe under global warming unless irrigation expands, fertilizer use increases, or new varieties are developed.

MEXICO'S CONTRIBUTION AND RESPONSE TO GLOBAL WARMING

Increases in the concentration of three major gases are currently thought to be changing the radiative properties of

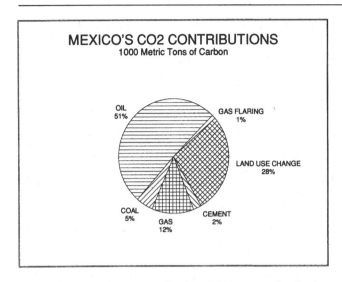

Figure 14: Mexico's CO₂ contributions (1000 tonnes of carbon).

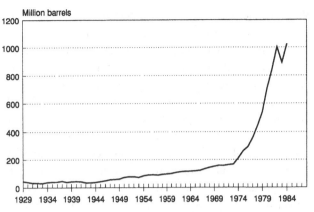

Figure 15: Petroleum production in Mexico.

the atmosphere in such a way that global temperatures will increase. According to the somewhat controversial World Resources Greenhouse Gas Index, Mexico is the world's thirteenth most significant greenhouse gas emitter (World Resources Institute 1990).

Most of Mexico's global warming contribution is in the form of CO_2 from fossil fuel combustion and land-use changes such as deforestation (Fig. 14). Methane is also significant. Of the fossil fuels, oil production is by far the greatest CO_2 contributor. The most significant methane sources are pipeline leaks and livestock. This snapshot of contemporary activities that produce greenhouse gases provides little insight into the history and possible future of greenhouse gas emissions.

Graphs based on historical statistics show some alarming trends. Petroleum production has grown dramatically in twentieth-century Mexico (Fig. 15). Wood production has also increased – just one indicator of the pressures that have caused widespread deforestation in Mexico. The dramatic growth of the livestock herd sets the context for contemporary methane emissions. Although most trends show a slowing of growth in the 1980s, it is unlikely that these trends will level off, or reverse, in the near future because of the economic and demographic pressures on Mexican resources.

Mexican population is still growing despite lower fertility rates, and at current rates may double in 35 years. Per capita consumption has also increased, with a slight fallback during the economic crisis in the early 1980s. Export growth, such as petroleum exports, is also driving resource consumption and agricultural expansion.

The forces and activities driving Mexico's greenhouse gas contributions are often seen as integral to industrial and agricultural development in Mexico. Slowing industrial growth, deforestation, and resource use in Mexico are diffi-

cult for a government burdened with debt, and for people striving for an improved standard of living.

And, of course, global warming may be seen as a low priority for a country facing more immediate economic, social and environmental crises. In the summer of 1991 we undertook a study of concern about global warming in Mexico, interviewing scientists, government officials and environmentalists. For most people in Mexico, local air and water pollution and deforestation are the priority issues. However, there is a long history of scientific and government interest in climate change in Mexico. For example, the meteorologists at the National University have their own climate model, and we found mention of global warming in the technical literature as far back as 1971.

The Mexican government demonstrated its commitment to reducing the risk of global change in 1987 when Mexico became the first country to sign the Montreal Protocol on the elimination of CFCs and the protection of the ozone layer. Mexico now intends to eliminate CFC use by 1997 and is actively negotiating for substitute technology. We were also told that the government views its tree planting and air pollution control policies as a response to the threat of global warming.

However, most people we interviewed were adamant about the unequal responsibility for the global warming problem. When the United States and other industrialized nations are disproportionately to blame for greenhouse emissions – the United States may produce 17% of the world's total according to the World Resources Institute (1990) – how can a less wealthy country like Mexico, trying to develop its economy and resources, be expected to reduce growth or really contribute very much to solving the global warming problem?

CONCLUSION

What are the policy options and sustainable development strategies for Mexico in the context of drought and future global warming? This chapter has established that climate variability and global warming have many serious impacts on Mexican agriculture and water resources. Mexico has been vulnerable to drought for centuries and the general direction of global change in Mexico is towards hotter, drier conditions. Global warming has significant implications for a wide range of agricultural, energy and conservation policies in Mexico and for many development projects planned or already under way. Any future hydroelectric or irrigation project should probably take account of possible declines in water supply and increased conflicts over access to water which may occur with global warming. The results also lend support to the calls for sustainable development, with less deforestation, greater care of soil and water resources, and stewardship of biodiversity.

REFERENCES

Adams, R. M., Rosenzweig, C., Peart, R. M., Ritchie, J. T. *et al.* 1990. Global climate change and US agriculture. *Nature* **345**:219–24.

Bach, W., Jung, H. J. and Knottenburg, H. 1985. *Modeling the Influence of Carbon Dioxide on the Global and Regional Climate: Methodology and Results*. Paderborn: Ferdinand Schoeningh.

Brown, B. 1985. A summary of late-quaternary pollen records from Mexico west of the isthmus of Tehuantepec. In *Pollen Records of Late-Quaternary North American Sediments*, ed. V. M. Bryant and R. J. Holloway.

Coll-Hurtado, A. 1982. *Es Mexico un pais agricola?: Un analisis geografico*. Mexico D. F. Siglo Veintiuno.

Dovring, F. 1970. Land reform and productivity in Mexico. *Land Economics* **46**:264–74.

Dunbier, R. 1968. *The Sonoran Desert: Its Geography, Economy and People*. Tucson: University of Arizona Press.

Ezcurra, E. and Montaña, C. 1990. Los recursos naturales renovables en el norte arido de Mexico. In *Medio Ambiente y Desarrollo en Mexico*, ed. E. Leff, pp. 297–327. Mexico: Porrua.

Florescano, E. 1969. *Precios del maiz y crisis agricolas en Mexico 1708–1810*. Mexico DF: Ediciones Era.

Florescano, E. 1980. Una historia olvidada: La sequia en Mexico. *Nexos* **32**:9–18.

Gleick, P. H. 1988. The effects of future climatic changes on international water resources: the Colorado River, the United States, and Mexico. *Policy Sciences* **21**:23–39.

Grotch, S. 1988. *Regional Intercomparison of GCM Predictions and Historical Climate Data*. Washington, DC: Department of Energy.

Hansen J., Fung, I., Lacis, A., *et al.* 1988. Global climate changes as forecast by the GISS three-dimensional climate model. *Journal of Geophysical Research* **93**:9341–64.

Hewitt de Alcantara, C. 1976. *Modernizing Mexican Agriculture*. Geneva: United Nations Institute for Research on Society and Development (UNIRSD).

IPCC 1990. Intergovernmental Panel on Climate Change. *Climate Change: The Scientific Assessment*. Cambridge: Cambridge University Press.

Jones, C. A. and Kiniry, J. R. 1986. *CERES-Maize: A Simulation Model of Maize Growth and Development*. College Station, Texas: A&M Press.

Kellogg, W. W. and Zhao, Z. 1988. Sensitivity of soil moisture to doubling of carbon dioxide in climate experiments. I: North America. *Journal of Climate* **1**:348–66.

Liverman, D. M. 1990. Vulnerability to drought in Mexico: the cases of Sonora and Puebla in 1970. *Annals of the Association of American Geographers* **80**(1):49–72.

Liverman, D. M. 1992a. The regional impacts of global warming in Mexico: uncertainty, vulnerability and response. In *The Regions and Global Warming*, ed. J. Schmandt and J. Clarkson, pp. 44–68. New York: Oxford University Press.

Liverman, D. M. 1992b. Global change and Mexico. *Earth and Mineral Sciences* **60**(4):71–6.

Liverman, D. M. and O'Brien, K. 1991. Global warming and climate change in Mexico. *Global Environmental Change* **1**:351–64.

Manabe, S. and Wetherald, R. T. 1986. Reduction in summer soil wetness induced by an increase in atmospheric carbon dioxide. *Science* **232**:626–8.

Manabe, S. and Wetherald, R. T. 1987. Large scale changes in soil wetness induced by an increase in atmospheric carbon dioxide. *Journal of the Atmospheric Sciences* **44**:1211–35.

Meehl, G. A. and Washington, W. H. 1988. A comparison of soil moisture sensitivity in two global climate models. *Journal of the Atmospheric Sciences* **45**:1476–92.

Metcalfe, S. E. 1987. Historical data and climate change in Mexico: a review. *Geographical Journal* **153**:211–22.

Mosiño, P. A. 1975. Los climas de la Republica Mexicana. In *El escenario geografico*, ed. Z. Cserna, pp. 57–171. Mexico: UNAM.

Mosiño, P. A. and Garcia. E. 1973. The climate of Mexico. In *The Climates of North America*, ed. A. Bryson and F. K. Hare, pp. 345–90. New York: Elsevier.

Nguyen, D. T. 1979. The effects of land reform on agricultural production, employment and income distribution: a statistical study of Mexican states 1959–1969. *Economic Journal* **89**:624–35.

Pearse, A. 1980. *Seeds of Plenty, Seeds of Change*. Oxford: Clarendon Press.

Reyes Castaneda, P. R. 1981. *Historia de la Agricultura: Informacion y Sintesis*. Mexico: AGT Editor, SA.

Schlesinger, M. E. and Mitchell, J. F. B. 1987. Model projections of the equilibrium response to increased carbon dioxide. *Review of Geophysics* **25**:760–98.

Schlesinger, M. E. and Zhao, Z. 1989. Seasonal climatic changes induced by doubled CO_2 as simulated by the OSU atmospheric GCM/mixed-layer ocean model. *Journal of Climate* **2**:459–95.

Servicio Meteorologico. 1976. *Normales Climatologicas*. Mexico: Servicio Meteorologico.

USEPA 1989. *The Potential Effects of Global Climate Change on the United States*. Washington, DC: USEPA (United States Environmental Protection Agency).

USEPA 1991. *Global Comparisons of Selected GCM Control Runs and Observed Climate Data*. Washington, DC: USEPA (United States Environmental Protection Agency).

Washington, W. M. and Meehl, G. A. 1986. Climate sensitivity due to increased CO_2: experiments with a coupled atmosphere and ocean GCM. *Climate Dynamics* **4**:1–38.

Wellhausen, E. 1976. The agriculture of Mexico. *Scientific American* **235**:128–50.

West, R. C. and Augelli, J. P. 1990. *Middle America*. Englewood Cliffs, NJ: Prentice-Hall.

Wilken, G. C. 1987. *The Good Farmers: Traditional Agricultural Resource Management in Mexico and Central America*. Berkeley: University of California Press.

Wilson, C. A. and Mitchell, J. F. B. 1987. A doubled CO_2 climate sensitivity experiment with a global climate model including a simple ocean. *Journal of Geophysical Research* **92**(D11):13315–43.

World Resources Institute 1990. *World Resources 1990–91*. Oxford: Oxford University Press.

3: The Impact of Climate Variation and Sustainable Development in the Sudano-Sahelian Region

FREDRICK JOSHUA WANG'ATI*

INTRODUCTION

The Sudano-Sahelian region is the strip of land south of the Sahara Desert extending across Africa from the Atlantic coast on the west to the Red Sea and the Indian Ocean to the east. The region becomes broader in the east to include the drier parts of Ethiopia, northern Uganda, Kenya and Tanzania (Fig. 1). The region embraces three ecological zones: the Saharo-Sahelian transition (arid), the Sahel (semi-arid) and the Sahelo-Sudanian (sub-humid).

The Saharo-Sahelian transition (arid) zone is characterized by low annual rainfall of 100–200 millimeters (mm) which also shows high inter-annual variability (50–100%). Climatologically, the humidity index, which is defined as the ratio of annual precipitation to potential evapotranspiration (P/ETP), is in the range 0.03–0.2. The vegetation is sparse and includes bushes and small woody, thorny or leafless scrub. In this zone, rainfed agriculture is not viable except in a few isolated flood-plain microenvironments. The main production enterprise is pastoralism.

The Sahel (semi-arid) and Sahelo-Sudanian (sub-humid) zones have better but still highly variable rainfall regimes (Fig. 2). The mean annual rainfall is in the range of 200–400 mm and 400–800 mm, respectively, with 25–50% inter-annual variability. The humidity index is in the range 0.2–0.5.

Rainfall distribution over most of the region is monomodal, occurring during the months of June, July and August, but the arid and semi-arid areas of Eastern Africa have two rainfall peaks, one in March and the other in May. It is this concentration of rainfall which supports crop production in

spite of the high evaporation rates. The most common soils are red or gray ferruginous leached soils with low natural fertility.

Climate variability, which affects mainly rainfall, has been a permanent feature in this region for hundreds of years. Recurrent droughts have had major impacts on the populations and their evolution of traditional survival mechanisms. These traditional strategies have not prevented major losses of life during particularly dry times, but numbers of people and livestock have built up again during the wetter periods.

Climate variation has therefore been one of the major factors contributing to environmental degradation and the socioeconomic process in the Sudano-Sahelian regions. It is thus imperative that patterns of climate variation are well understood as a basis for the formulation of more sustainable developmental policies and strategies, taking into consideration the changing socioeconomic circumstances which face the populations in the region.

Population and migration

The population of the 22 countries of the Sudano-Sahelian region is estimated to be 347 million and is increasing at a rate of 2.7–2.8% per annum (Hammond and Paden 1990). There has been considerable internal and external migration of people, especially from rural to urban areas, but also to other African countries, mainly as a result of pressures caused by climatic variations, environmental degradation and population increases.

Economic activities

Of the 22 countries in the region, 18 are designated among the world's poorest. Apart from a few mining activities, especially in the western parts (Mali, Mauritania and Niger), there are few major industrial enterprises in the region and

* This report was prepared under the auspices of the United Nations Sahelian Organization (UNSO) by Fredrick Wang'ati (Consultant on Agriculture, Research and Development, Kenya) with contributions from Dr Gideon-Cyrus Makau Mutiso (Socio-Economist, Kenya) and Kaliba Konare (Director of Meteorological Services, Mali).

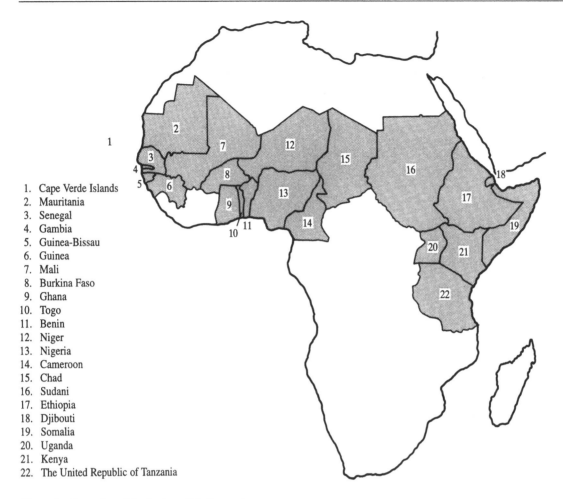

1. Cape Verde Islands
2. Mauritania
3. Senegal
4. Gambia
5. Guinea-Bissau
6. Guinea
7. Mali
8. Burkina Faso
9. Ghana
10. Togo
11. Benin
12. Niger
13. Nigeria
14. Cameroon
15. Chad
16. Sudani
17. Ethiopia
18. Djibouti
19. Somalia
20. Uganda
21. Kenya
22. The United Republic of Tanzania

Figure 1: Countries of the Sudano-Sahelian region.

the population is dependent mainly on subsistence agriculture and pastoralism. However, intensive crop production is practiced in the Nile, Niger and Senegal river valleys which produce food as well as industrial crops such as cotton.

Economic, social and environmental sustainability

As a result of the poor economic base, education and health facilities are still very limited and the region suffers from a severe shortage of trained manpower. The region also is composed of a fragile ecosystem where a delicate balance has to be maintained between conservation and exploitation of the vegetation, water and soil resources. Sustainability of both the environment and socioeconomic development activities should therefore be a major concern of the national governments, donors and other development agencies.

The problem of sustainability was clearly demonstrated during the severe drought which greatly affected the countries of Mauritania, Senegal, Mali, Chad, Burkina Faso, Sudan, Ethiopia and Somalia during the 1970s and early 1980s. The local systems were clearly unable to cope and even

massive external assistance in the form of emergency famine relief and health support could not prevent huge losses of life.

Potentialities and problems

In spite of their present economic situation, most countries of the Sudano-Sahelian region have a large reservoir of untapped resources of land, water, sunshine, wind and geothermal energy, unexplored mineral resources, and human labor. Many problems relating to social organization and stability, availability of capital, priority setting in planning, and availability of appropriate technology have, however, prevented the optimum development of these resources.

TRENDS IN KEY VARIABLES

Natural resources

Climate

Rainfall is the most variable and critical climatic parameter in the Sudano-Sahelian region. According to the World

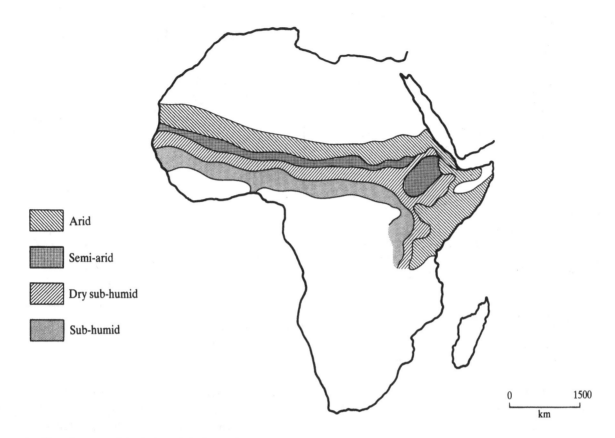

Arid

Semi-arid

Dry sub-humid

Sub-humid

0   1500
        km

Figure 2: Climatic zones of the Sudano-Sahelian region.

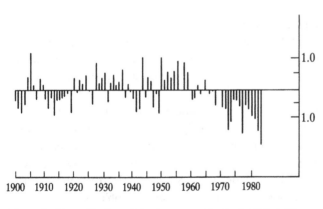

Figure 3: Normalized precipitation anomalies in the Sahel, West Africa. (Source: WMO/UNEP (1991).)

Meteorological Organization/World Climate Program (WMO/WCP) publication on the global climate system, one of the striking features of the rainfall pattern in Sub-Saharan Africa since the early 1960s is that each dry episode has been more extreme than the previous one, giving the impression of a general downward trend in rainfall (Fig. 3).

There is, therefore, historical evidence that severe droughts have been part of climatic variations in the past. These droughts occur on two time scales – short term, lasting only 1 or 2 years, and long term, lasting over 4 years – but there is no clear pattern which can be used for their prediction. There is also mounting evidence that such droughts are not caused by local human activity but are most likely of global origin linked to large-scale oceanographic and atmospheric phenomena (e.g. El Niño Southern Oscillation: ENSO). It is, however, possible that droughts in this region could be accentuated to some extent due to rapid deforestation and desiccation of the vegetation and the resultant changes in energy balance. It has therefore been difficult to state categorically the extent to which the current downward trend will persist or recur in the future (Nicholson 1989), or the long-term effects it would have on the ecology.

A good example of the trend in climatic variation is the behavior of the rainfall regime in the Sudan. Analysis of this parameter (Hulme 1989) shows that during the period 1965–85, rainfall declined by 20% throughout Sudan, resulting in contraction of the wet season in the semi-arid central Sudan by about 3 weeks and a southward displacement of rainfall zones by 50–100 kilometers (km). Similar shifts in isohyets (Fig. 4) have been observed in the Gambia–Senegal–Mauritania region (Republic of Senegal 1991).

Although large quantities of climatological data have been recorded and in some cases stored in computerized formats within the region (e.g. AGRHYMET Centre, Niamey), these data have not been adequately analyzed due to the lack of

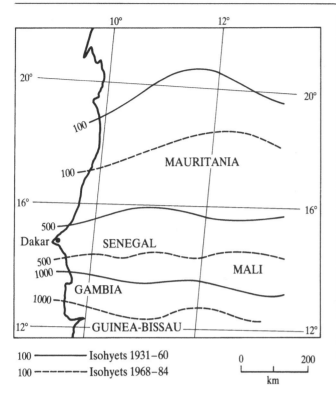

100 ——————— Isohyets 1931–60

100 – – – – – – – Isohyets 1968–84

0   200
                km

Figure 4: Movement of isohyets in the Mauritania–Senegal–Gambia region, 1968–84. (Source: Republic of Senegal 1991.)

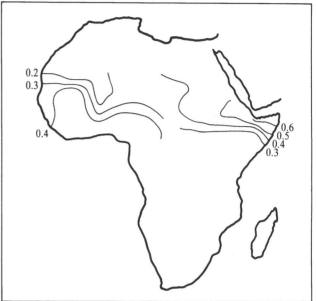

Figure 5: Probability of a decrease in precipitation in June, July and August in the Sudano-Sahelian region.

financial and trained human resources. It is therefore difficult at present to make detailed inferences on long-term regional climate trends, except for the general picture emerging from studies of global climate change. One such analysis has been carried out by Downing and Parry (1991). Their findings (Fig. 5) suggest that countries such as Djibouti, Somalia and Ethiopia will have a higher probability of reduction of rainfall during the months of June, July and August (which constitute the rainy season within the region) compared with their counterparts in the western part of the region.

The full impact of global warming, which is associated with build up of greenhouse gases in the upper atmosphere, is still difficult to predict. If, as currently estimated, the surface temperatures rise by 1 °C in the next 30 years, increases in potential evapotranspiration and unknown changes in rainfall regimes are bound to occur. These changes may or may not compensate one another.

Water resources
The major water systems in the region are the rivers Nile, Niger, Senegal, Gambia and Chari, and lakes Chad and Turkana. Groundwater is also tapped extensively, but it is often of low quality due to the high concentration of dissolved salts. Variations in rainfall have a direct influence

on these water resources, resulting in severe floods such as those which occurred on the Nile recently, and drying of lakes as has been observed in Lake Chad and Lake Turkana. Trends in these resources are, however, difficult to quantify due to the low level of physical monitoring.

Land resources
Land degradation is believed to have increased during the past 20 years due to climatic variations and increasing pressure from both human and livestock populations, which tends to reduce the regenerative capacity of the natural vegetation. Few systematic and comprehensive surveys of land degradation, usually covering only limited areas of developmental interest, have been carried out in the region. The data in Tables 1 and 2, derived from Global Assessment of Soil Degradation (GLASOD) data (UNEP 1991), provide some indication of the extent of the problems.

Causative factors for soil degradation are indicated as: agriculture, 24.6%; overexploitation, 12.7%; deforestation, 13.5%; overgrazing, 49.2%; bio-industrial activities, 0% (UNEP 1991). The influence of climate variations is implicit in these factors.

Although these figures refer to the whole of Africa, there is no doubt that the majority of the badly degraded lands are located in the Sudano-Sahelian region. Prolonged drought will have the effect of delayed or reduced re-establishment of vegetation cover, thus reducing livestock populations and increasing soil degradation due to depletion of organic matter. These data should, however, be treated with caution,

Table 1. *Degree of soil degradation in various bioclimatic zones in Africa (millions of hectares)*

Degree	Bioclimatic zone		
	Dry sub-humid	Semi-arid	Arid
None	231.4	404.3	331.0
Light	13.8	23.7	80.5
Moderate	11.4	46.2	69.6
Strong	11.3	38.1	21.3
Extreme	0.7	1.5	1.2
Total	268.6	513.8	503.6

Source: UNEP (1991).

Table 2. *Types of soil degradation in various bioclimatic zones in Africa (millions of hectares)*

Type	Bioclimatic zone		
	Dry sub-humid	Semi-arid	Arid
None	231.4	404.3	331.0
Water	25.1	59.2	34.8
Wind	1.6	30.7	127.5
Chemical	7.7	11.3	7.6
Physical	2.9	8.4	2.6
Total	268.7	513.8	503.5

Source: UNEP (1991).

both since they are based on informed opinion rather than scientific measurements and because of the broad scale on which generalizations have been made (UNSO 1992).

The term 'desertification' has been defined by The United Nations Environment Programme (UNEP) as land degradation in arid, semi-arid and sub-humid areas resulting mainly from human impact. This process is therefore not confined to the Sudano-Sahel region but is taking place in areas where the ecological balance has been badly damaged or destroyed.

Desertification in the Sudano-Sahelian region has attracted more attention since it has appeared as a non-reversible expansion of the Sahara Desert southward. Lamprey (1988), for example, concluded that the southern boundary of desert vegetation in western Sudan had moved southward by 60–100 km between 1958 and 1975, an alarming rate of 5½ km per year. However, field studies by Hellden (1984) in the same area found no evidence of such a change. The theory of 'advancing Sahara' has been shown to be faulty; it would be more correct to consider such changes as oscillations of the Sahara boundary which are dependent on periodic fluctuations in rainfall.

It is nevertheless important to point out that although some of the badly degraded areas may appear to be recovering, these gains may well be reversed if adequate measures are not taken to restore reasonable ecological balance.

Forest and energy resources
Information on status and trends in forest and energy resources is scanty, but some of the data available indicate a steady decline in forest resources by approximately 0.5–3% per annum during the 1980s. The main reasons for deforestation are expansion of cultivation, fuel wood and grazing. It would appear that unless new and renewable sources of energy can be found (e.g. through increased development of geothermal energy), deforestation will increase and the forest resources in the region may be largely depleted within the next 50 years.

Land use

Traditional land use in the Sudano-Sahelian region is comprised of shifting cultivation, smallholder agriculture and nomadic pastoralism. Although no major changes in land use are expected to occur in the foreseeable future, there are already signs that sedentary agriculture will increase as a result of population pressure. Some large-scale mechanized farming has been tried on marginal lands, especially in the Sudan. But it may be too early to predict future trends in these activities, since there are already signs that such developments may be abandoned if the cost of inputs relative to commodity prices in the world market makes the enterprises unprofitable. If lower rainfall persists, major changes in land use are inevitable as previously cropped land is abandoned and taken over by pastoralists. Even the latter is dependent on surface water reservoirs which are replenished by seasonal rainfall. Land tenure is another largely unresolved and highly political issue which will influence trends in land use.

Food production and balance

One way of showing trends in food production and balance is to use an index representing the average annual quantity of food produced per capita in relation to that produced in the indexed year, and the ratio of food imports to the food available for internal distribution. Table 3 shows these parameters for a number of countries in the region.

These data indicate a stagnation or a decline in food production per capita and an increasing dependency on food imports during the 1980s. The figures do not reveal, however, the well-known inequities in access to food among various

Table 3. *Trends in food production and import dependency in some Sudano-Sahelian countries*

	Food production index per capita (1979/81 = 100)	Food import dependency ratio (%)	
	1986–8	1969–71	1986–8
Senegal	106.0	31.7	30.2
Ethiopia	89.0	—	—
Sudan	89.0	9.8	14.5
Somalia	100.0	13.1	23.7
Chad	103.0	3.8	4.5
Burkina Faso	116.0	3.7	9.6
Niger	83.0	2.0	7.4
Mali	97.0	6.0	9.6
Gambia	95.0	19.8	55.4
Average	97.5	11.2	19.4

Source: UNDP (1991).

Table 4. *Economic trends in some Sudano-Sahelian countries*

	Annual % increase		Annual % growth rate in GNP per capita	
	Arable land 1965–85	Livestock 1975–85	1965–80	1980–8
Senegal	0.37	1.3	−0.5	−0.3
Ethiopia	0.06	0.1	0.4	−1.4
Sudan	0.10	3.7	0.8	−4.2
Somalia	0.02	0.7	−0.1	−2.2
Chad	0.03	2.5	−1.9	—
Burkina Faso	0.21	2.8	1.7	2.4
Niger	0.45	3.7	−2.5	−4.2
Mali	0.07	1.2	2.1	0.4

Source: UNDP (1991).

social sectors of the populations, which are bound to worsen with increases in food import dependency.

Economic output

Table 4 shows trends in the national economies of some of the countries in the region. The figures indicate that the economies are in decline.

Population and migration

Table 5 illustrates trends in population growth in some of the countries most affected by climate variation in the region. These figures are projections based on the 'medium variant' derived from geometric calculations, which are considered more reasonable than the 'high variant' projections derived from exponential calculations. According to these projections, the population is expected to double between the years 2000 and 2050, which will impose even greater pressure on the environment.

Seasonal migration has been a traditional feature of the Sahel region for hundreds of years as a means of escaping the hazards of the harsh environment and a search for employment. This practice has, however, intensified in recent years due to persistent drought and ecological degradation. An important feature of this migration is that it involves mainly young males, thus reducing the capacity of the remaining population to work on the land. Although accurate data are difficult to obtain, it has been estimated that the rate of migration from destitute regions of the Sahel is among the highest in the world (David and Myers 1991). There are

indications that this trend will continue with the encouragement of the governments as a means of generating foreign exchange income through remittances.

A disturbing trend in migration is the movement of entire sections of populations as a result of prolonged drought. This type of migration is on the increase and has created a new problem of 'environmental refugees' in the region. In the absence of suitable areas for new settlement, large numbers of such people are now to be found camping on the peripheries of major towns and other urban areas, where they seek relief and employment. Many of these people hope to return to their original homes in the future and therefore do not have a long-term interest in the new environment. These settlements are therefore likely to accelerate the rate of land degradation in the region.

Urbanization

The proportion of the population living in urban areas is increasing and is expected to double during the period from 2000 to 2025 (Table 6). While some of this increase will be from natural growth, the majority of the new urban dwellers will be immigrants from rural areas. Indications are that the high rate of rural–urban migration will continue, or even accelerate if rural incomes continue to fall as a result of environmental degradation, unemployment and lack of social amenities commensurate with improved literacy.

Education and health

Social progress within a community can be measured by the degree to which the three problems of underdevelopment, i.e.

Table 5. *Population trends (medium variant) in some Sudano-Sahelian countries (thousands)*

	1950	1960	1970	1980	1990	2000	2025
Djibouti	60	80	168	304	409	552	1 094
Somalia	2 423	2 935	3 668	5 345	7 497	9 736	18 701
Chad	2 658	3 064	3 652	4 477	5 678	7 337	13 245
Sudan	9 190	11 165	13 859	18 681	25 203	33 625	59 605
Mali	3 520	4 375	5 484	6 863	9 214	12 685	24 774
Niger	2 400	3 028	4 165	5 586	7 731	10 752	21 482
Senegal	2 500	3 187	4 158	5 538	7 327	9 716	16 988
Burkina Faso	3 654	4 452	5 550	6 957	8 996	12 092	23 710
Ethiopia	19 573	24 191	30 623	38 750	49 240	66 364	126 618

Source: UNFPA (1991).

Table 6. *Percentage growth of the population in urban areas of some Sudano-Sahelian countries*

	1950	1960	1970	1980	1990	2000	2025
Djibouti	41.0	49.6	62.0	73.7	80.7	84.3	89.7
Ethiopia	4.6	6.4	8.6	10.5	12.9	16.8	33.8
Somalia	12.7	17.3	22.7	28.9	36.4	44.3	62.5
Chad	4.3	7.0	11.3	20.2	29.5	38.8	57.9
Sudan	6.3	10.3	16.4	19.7	22.0	26.5	45.5
Mali	8.5	11.1	14.3	17.3	19.2	23.2	41.7
Niger	4.9	5.8	8.5	10.3	13.2	19.5	26.8
Burkina Faso	3.8	4.7	5.7	7.0	9.0	12.4	27.3
Senegal	30.5	31.9	33.4	34.9	38.4	44.5	62.6

Source: UNFPA (1991).

Table 7. *Percentages of school-age children enrolled in schools in some Sudano-Sahelian countries*

	Percentage age-group enrolled in schools			
	Primary		Secondary	
	1965	1987	1965	1987
Ethiopia	11	37	2	15
Chad	34	51	1	6
Somalia	10	15	2	9
Burkina Faso	12	32	1	6
Mali	24	23	4	6
Niger	11	29	1	6
Sudan	29	49	4	20
Senegal	40	60	7	15

Source: World Bank (1990).

poverty, lack of education and poor health, have been alleviated. It is, however, difficult to measure these parameters directly and a common solution is to use indicators such as income per capita and its distribution, the number of school-age children enrolled in schools, access to food and health care and ultimately the average life expectancy at birth. Trends in these parameters are summarized in Tables 7 and 8 for some of the countries in the heart of the Sudano-Sahelian region. It is, however, important to point out that accurate censuses are rare in the region and the figures presented should be regarded as estimates.

It is very encouraging to see the progress achieved in education in all the countries of the region since they attained independence. This progress is, however, seriously threatened by the deteriorating economic trends, which are caused to a large extent by recent droughts, heavy external debt and political instability. One area which is already affected is access to health services, which does not keep up with population increases.

Table 8. *Access to health care and life expectancy in some Sudano-Sahelian countries*

	Population per physician		Av. life expectancy (yrs)	
	1965	1984	1977	1988
Senegal	19 490	13 060	42	48
Ethiopia	70 190	78 970	39	47
Sudan	23 500	10 100	46	50
Somalia	36 840	16 080	43	47
Chad	72 480	38 360	43	46
Burkina Faso	73 960	57 220	42	47
Niger	65 540	39 730	42	45
Mali	51 510	25 390	42	47

Source: World Bank (1989, 1990).

Sustainability

It is evident from the foregoing trends that there is a serious problem of sustainability of the ecological balance, the natural resources and the development and welfare of the nations and populations of the Sudano-Sahelian region. While climate variations are a natural feature of the region, a great deal could be done to reduce their negative impacts on the environment and to enable the populations to cope better with the inevitable short-term and long-term climatic changes. Some of the options available and strategies which could be adopted are discussed below.

CLIMATE AND SOCIETY

Past and present impacts of climatic variability and societal responses

Impacts of climatic variability are felt most at the national level. Societal responses have therefore been most intense at the national level, but cooperation and assistance has also developed at regional and global levels. Each of these levels will therefore be discussed separately.

Impacts and responses at global level
Climatic variations have stimulated considerable activities at the global level. The most notable are the World Meteorological Organization (WMO) World Climate Program, WMO/UNEP Intergovernmental Panel on Climate Change (IPCC) and the other numerous international activities leading to UNCED. The United Nations Sudano-Sahelian Office (UNSO) was also established by the UN Secretary General in 1973 following the disastrous drought that struck the Western Sahel in the late 1960s and early 1970s, to assist in the rehabilitation of the drought-stricken countries of this region. Through these activities, a large effort has been deployed to gain a better understanding of climatic variations and their impacts on the ecology and the extent to which such variations can be predicted. As a result of these studies, the low level of economic and technological development and international imbalances are emerging as significant factors in the ability of developing countries to cope with climatic variability. The influence of past and present global response on the actual capacity of the region to cope with climatic variations is, however, yet to be felt, and only time will tell whether a more meaningful and sustainable global response will emerge from these efforts.

Impact and responses at regional level
The most visible impact of climate variability at the regional level has been the huge interstate migrations of destitute populations in search of famine relief. This problem affects most of the countries in the region, especially those bordering the Sahara Desert, and has been accentuated in several cases by internal political instabilities. Such migrants have caused considerable political tension between countries and economic hardship to the host countries.

At regional levels, countries have responded by establishing regional organizations, in particular the Permanent Inter-State Committee for Drought Control in the Sahel (CILSS) and the Intergovernmental Authority on Drought and Development (IGADD), whose mandates include drought and development.

CILSS was established in 1973 and its membership comprises nine countries: Burkina Faso, Cape Verde, Gambia, Guinea-Bissau, Mali, Mauritania, Niger, Senegal and Chad. The main aspects of the work of CILSS are:

- the promotion of programs for food self-sufficiency in the region;
- the promotion and coordination of development programs within the region;
- an attempt to sensitize donors to the priority needs of the region.

The major programs undertaken by CILSS include:

- establishment of the Sahelian Network of Documentation;
- pest and disease control unit;
- establishment of two regional programs: Centre d'Agrométéorologie et d'Hydrologie (AGRHYMET) and Institut du Sahel (INSAH) and nine national projects;
- a permanent Diagnostics Project (DIAPER);
- a plant breeding program (millet and sorghum);
- an energy program (gas and solar).

The international response to these programs was the establishment of a liaison body (Club du Sahel) to coordinate donor assistance.

The main impact of these programs has been to increase the numbers of trained specialists in various fields of development and the implementation of over 600 development projects in the member countries.

IGADD was established in 1986 and its membership is comprised of Djibouti, Ethiopia, Kenya, Somalia, Sudan and Uganda. The chief aim of IGADD is to enable member states to pool and coordinate their developmental efforts in combating drought and its effects on the environment, food, security and socioeconomic development in the region in a sustainable manner. Long-term programs have therefore been focused on:

- Problems of food security including food policy management, protection and increase in food production and efficiency of marketing. An early-warning system for food security was launched in 1989.
- Desertification control. This involves use of assessment techniques including remote sensing, optimal and sustain-

able use of natural resources, rehabilitation and conservation of the environment.

IGADD is, however, a relatively young organization and is still in the process of formulating its programs. It has nevertheless stimulated the harmonization of environmentally related development activities in the region and has, through donor-assisted projects, collected much useful information on ecological characteristics and changes in the subregion.

A regional Drought Monitoring Center (DMC) for eastern and southern Africa was also established in 1989. The DMC is a UNDP project implemented by WMO and has operational bases in Nairobi and Harare. The DMC is already issuing monthly bulletins one month in advance on drought severity, rainfall anomalies and weather outlook.

Establishment of these organizations has also been stimulated by the hope and expectation that they will be in a better position to attract technical and development assistance from the developed countries and United Nations agencies. This is already happening, but not at the level of original expectations.

Impacts and responses at national level
It would appear that in most countries of the Sudano-Sahel region the major impacts of climate variations have been droughts, land degradation and desertification. Societal response has been to evolve various strategies for adapting, including migration as a last resort. According to the report 'Assessment of desertification and drought in the Sudano-Sahelian Region 1985–1991' prepared by UNSO for UNCED, and a number of other national reports prepared for UNCED, governments have responded with programs to increase the level of public awareness of causes and dangers of land degradation. Success is being achieved through mobilization of populations to counteract drought and desertification through various programs and projects. National early-warning systems and environmental monitoring systems have been established in a number of countries of the region and integrated development projects now include participation by meteorological services. At the community level, response strategies include increased use of water-harvesting techniques to safeguard both crop and livestock production, and initiation of small-scale irrigation projects. Other activities include dryland agriculture, range management, deforestation control, soil protection and sand-dune fixation, water resource management and development.

Impacts of climatic variability and societal responses at the national level do, however, vary from country to country and it is difficult to provide detailed accounts of impacts and responses for all countries of the region. However, a brief account of some characteristics, including history and exper-

Table 9. *Area by agro-ecological zones in Kenya*

		% R/Eo	Area (km²)	% country area
Zone IV	Semi-humid	40–50	27 000	5
Zone V	Semi-arid	25–40	87 000	15
Zone VI	Arid	15–20	126 000	22
Zone VII	Very arid	15	226 000	46
Total			506 000	88

Note: R/Eo, ratio of annual rainfall (R) to evapotranspiration (Eo).
Source: Jaetzold and Schmidt (1983).

iences with impacts of climate variations, migration, settlement and development, in the arid and semi-arid lands (ASALs) of Kenya, which are similar to many parts of the Sudano-Sahelian region, could provide useful insights.

Permanent migration and settlement in drylands: experiences in the ASALs of Kenya

Historical. Over the past 90 years, many farmers have moved from the more humid zones to the less humid zones. As illustrated in Table 9, zones IV and V account for 20% of the country and are significant for crops and livestock production. Zone VI forms 22% and zone VII accounts for 46% of the country. Therefore, 88% of Kenya has less than 800 mm annual rainfall. The development and reclamation of land in these zones is a clear national priority because of the exploding population and the need to feed that proportion of the population that resides in the ASALs with food produced there.

Historically, zones IV, V and VI have received a migrating population from the more humid zones. In the perspective of centuries, it is doubtful whether there were permanent settlements outside zones II and III until the early eighteenth century, as most oral traditions attest. It has been learned from oral traditions that ASAL production was integrated into the hill/mountain-based homesteads through hunting and livestock keeping in *syengo* (cattle camps) in the dry plains. The issue of continuous use of the fragile ASAL ecosystem did not arise for the *syengo* because of the constant relocation of the bases; thus there was intermittent use of range resources.

The *syengo* was not only for herding. From it came major social institutions for deployment of labor, the distribution system of livestock and grain consumables, the scattering of livestock resources so as to escape drought and disease, and the then dominant land-holding form which assured that every family owned both mountain and plain land, as is

Table 10. *Livestock population in Kenya, 1987 (thousands)*

	Beef cattle	Dairy cattle	Sheep	Goats	Camels	Donkeys
Total ASAL	5761	715	4144	7283	956	249
Total non-ASAL	3310	2287	2300	1245		
ASAL as % of total	64	24	64	85	100	100

Source: GOK (1989).

Table 11. *Average land-holding (hectares per person) in selected ASAL districts of Kenya*

District	1969	1979	1989
Narok	7.32	4.30	2.66
Lamu	3.36	1.76	0.98
Laikipia	2.09	1.03	0.55
Kitui	0.89	0.66	0.50
Kwale	0.79	0.57	0.42
Embu	0.58	0.35	0.28
Kilifi	0.53	0.38	0.28
Taita	0.45	0.34	0.26
Machakos	0.40	0.28	0.20

Source: Livingstone (1989), quoted in GOK (1990).

found in the traditions of many Kenyan peoples. These economic and ecological adaptation mechanisms, encapsulated in the institution of the *syengo*, have been marginalized by the population growth of the past ninety years mainly as a result of permanent migration from the more humid areas.

ASAL production. The ASALs produce the bulk of the meat products in Kenya (Table 10). Although ASALs produce subsistence crops for their population, national food statistics do not provide a coherent picture of the contribution of ASALs to crop production. However, the bulk of the bean, cowpeas, pigeon peas, simsim, millet and sorghum, crops which form a major pillar of national food consumption, are produced in the ASALs (GOK 1990). However, yields per hectare are generally low and vary between years, seasons and inter-cropping situations.

ASAL land-holding. The fact that land is becoming short in the ASALs is well demonstrated in Table 11, which shows average land-holdings in selected districts.

This land shrinkage presents tremendous challenges for sustainable development and intensification of ASAL production. The key issues in any intensification are maintenance of soil fertility, labor-saving tillage and handling

equipment and water-harvesting for production and, perhaps most complex, the integration of crops and livestock production so as to enrich the land.

The objectives of the ASAL programs were ranked by the government of Kenya in 1979 as:

(1) Development of human resources.
(2) Exploitation of productive potential.
(3) Resource conservation.
(4) Integration with the national economy.

The effect of the proclamation of the 1979 ASAL Strategy at the meta-policy level was to create a framework for channelling resources to areas which would not receive them under normal economic policy concerns. In particular, the project selected criteria which preferred projects with the highest rates of return in the short term. The government of Kenya was sending a clear message to the donors that the areas deserved development in their own right.

Under the 1979 ASAL Strategy, the main program approach was to be integrated development, which by implication was to be area based. The level of government which was relevant, therefore, was the district, but this logical framework did not always work because some donors have operated in ASAL districts at levels lower than that of the district.

This ASAL district based approach to development adhered precisely to notions of decentralization of government operations (project identification and planning, budgeting and finance operations) which were initiated in 1966 within the civil service, but which did not get clear backing from the political arena until much later. It was, however, from the political arena that the momentum for the District Focus Strategy for Rural Development (DF) was eventually generated.

In recognition of the potential importance of ASALs, and the complexity of development processes in these areas, the Kenyan government created, in May 1989, the Ministry of Reclamation and Development of Arid and Semi-arid and Wastelands (MRDASW). The new Ministry is expected to provide enhanced capability within the Kenyan government to address the special constraints and development possibilities of ASALs.

The primary objectives of the 'second-generation' strategy for arid and semi-arid lands reclamation and development, as set forth in a document under consideration, are to:

● develop ASAL capacity for income generation, employment creation, and the attainment of food security;
● reclaim and protect the fragile ASAL ecologies;
● improve the quality of life for the present and future generations of ASAL inhabitants on a sustainable basis.

The key programs of this strategy are crop development, livestock development, water development, natural-resource conservation, infrastructure, off-farm employment including enterprise development, and health and social services development.

The Kenyan lesson. The Kenyan ASAL experience is that of population moving into and within ASALs as a result of population pressure. Although migrants faced many initial difficulties and conflicts of interest, they have been able to settle and evolve a form of existence which appears to be sustainable, at least in the short to medium term. A major factor in this success has been the recognition by the government of the special needs of such regions and a constant review of development policies and strategies for their applicability and sustainability. While these strategies have helped the population to cope, it is recognized that there is a limit to the size of the population that the ASALs can carry on the basis of the land resources. Future responses will therefore depend on the availability of new technologies for increasing land productivity, success of the family planning programs now under way and migrations out of the ASALs.

Possible future impacts of climatic variability and capacities to cope

Climatic variability is likely to remain a major feature of the Sudano-Sahelian environment, and there is clearly very little man can do to change this situation. The future impacts of these variations will, however, depend on the capacities of the population to cope, especially the state of preparedness at national and individual community levels.

Impacts assuming present climate variability

The Sudano-Sahelian region is a fragile ecosystem, the capacity of which to withstand the normal climatic variations is critically dependent on the pressures exerted on the land and vegetation resources by both human and livestock populations. The trends emerging for these parameters, however, suggest that pressure on the environment (environmental stress) will increase rather than decrease in the foreseeable future, even if one assumes no change in the nature, frequency and magnitude of climate variations. Two of the main causes of this pressure are rapid population increase and reduced mobility across borders within the region.

Before the countries of the region attained independence, the largely nomadic populations were able to move their livestock freely over long distances in 'transhumance' systems following the seasonal distribution of rainfall and the resulting pastures. The 'transhumance' system practiced between the Gourma and the seasonally flooded Niger delta in Mali, and similar systems operating between Senegal and Mauritania, are examples of this movement. These pastoralists were also able to depend on traditional and cultural agreements between themselves and the farmers in the more fertile regions for exchange of grain and fodder for livestock products, thus avoiding the necessity to cultivate marginal lands. Similar arrangements enabled the farmers in the fertile, but more crowded regions to put their livestock under the care of the nomadic pastoralists such as the Fulani.

These systems have, however, come under severe stress due to prolonged droughts, imposition of state boundaries, interstate conflicts and land-tenure restrictions. As a result, large numbers of nomadic pastoralists have lost their livelihood and are being compelled by circumstances to adopt a sedentary life in marginal lands. Under these circumstnces, the land will be unable to recover from even mild climatic variations. Intense rainstorms falling on ground with scanty, if any vegetation cover will result in flash floods and soil erosion which will reduce even further the carrying capacity of the land. The already serious problem of environmental refugees and rural–urban migration will accelerate, increasing pressure on the already inadequate infrastructures and social services. Interstate migration of destitute populations will also increase, thus aggravating economic, social and environmental stress in host countries.

Impacts assuming climate change

A major global climate change could have an even more profound impact on the region. The IPCC (1991) study showed that if global emission of greenhouse gases continues at the current levels, the earth will warm up by 1–3 °C by the year 2030. Such a rise in temperature is expected to cause more intensive weather activity, resulting in more severe and frequent floods and droughts. Some regions of the world are also expected to dry out, while others could become too wet to allow farming activities. Present predictions of future changes in precipitation and regional distributions of climate change are hardly trustworthy. They must be verified against the changes that are taking place, and will take place in the next decade (Bolin 1989). In either case, climate change could be expected to have an impact on the range of crop, vegetation and other wildlife species which will thrive in the new environment.

The IPCC (1991) impacts assessment also predicts a sea level rise of 0.3–0.5 meters by the year 2050 and about 1 meter by the year 2100. Although most of the countries of the region are far from coastlines, such changes could have serious effects on beaches and tourism, especially on the Indian Ocean Coast. In addition, there could be huge losses of farming land in countries such as Gambia, Guinea Bissau and Senegal.

There are bound to be time lags of uncertain duration between changes in the physical (climatic) and ecological effects, and the resultant socioeconomic consequences. Changes will not necessarily come at a steady pace, and surprises cannot be ruled out (Tegart *et al.* 1990). Depending on the time scale in which such changes occur, they could

Table 12. *Population carrying capacity (millions) of some Sudano-Sahelian countries under different levels of inputs*

Country	Projected population year 2000	Population potential and levels of inputs		
		Low	Medium	High
Burkina Faso	11.8	6.0	26.7	135.2
Chad	8.0	14.2	70.1	306.4
Ethiopia	55.3	20.1	70.8	280.4
Mali	11.6	10.5	37.3	171.3
Niger	9.7	1.4	3.1	43.4
Senegal	9.7	7.6	22.3	105.8
Somalia	6.3	2.1	3.2	7.7
Sudan	31.3	80.6	259.4	1054.7

Source: FAO/ UNFPA/IIASA (1984).

significantly reduce the capacity of the majority of the populations to adapt. The shorter the time lag, the lower the ability to cope and the greater the impacts will be.

ENVIRONMENTAL STRESS AND DESERTIFICATION IN THE FUTURE

Environmental stress may be viewed in three distinct processes – drought, desiccation and degradation (desertification) – the effects of which are to reduce sustainability of the ecosystem and capacity of the environment to support human activities. Climate has a direct influence on all these processes which could take place with or without human intervention. Prospects for the future therefore depend mainly on trends in climate, but also to a significant extent on the degree to which human activities affect the natural capacity of the ecosystem to withstand climate variability.

Land carrying capacity, however, is a function of natural resources, mainly soil and water, and the level of inputs of technologies, labor, capital, etc., available to optimize outputs in a sustainable way. Although such information is crucial for overall planning and assessment of environmental stress and its impacts, such studies are extremely complex and comprehensive data on a large scale are difficult to obtain. Data available (Table 12) suggest that carrying capacity will be exceeded, in some cases by over 100%, if levels of inputs remain low. A wide range of parameters and assumptions have been used in arriving at the projections, and more information is available through the source.

The main inputs considered are those relating to the enhancement of the agricultural base, namely soil fertility, crop varieties and farming technology. In drier zones, the rate of mineralization of organic matter exceeds the rate of

decomposition, and the humus content falls over the long term. The effects of long-term climate change on crop yields may therefore be quite different from those of inter-annual climate variations of similar magnitude (Parry *et al.* 1988). Indications are, however, that with the worsening terms of trade and decreasing per capita incomes, prospects for raising the level of inputs are poor. Land degradation and desertification are therefore bound to increase as soils are 'mined' beyond their capacity to regenerate fertility naturally, and the region will continue to experience increasing food deficits and economic hardship. Carrying capacity should, however, not be based entirely on primary production. Market economic systems have a major influence and there is a need to maintain up-to-date information on these trends.

Another serious environmental problem which influences areas of the Sahel bordering the desert is the encroachment of sand on pastures, arable land and human settlements. This problem has affected large areas of pastoral land in Mauritania and is threatening the western boundary of the Gezira irrigation scheme which is the lifeline of the Sudan. Sand encroachment is bound to worsen with increased sedentarization of nomadic pastoralists and reduction of the vegetative cover. Control measures involve establishment of windbreaks and shelterbelts, a process which will become even more difficult if the rainfall decreases.

Environmental stress therefore makes prospects for life in many parts of the Sudano-Sahelian region uncertain unless urgent and coordinated efforts are made at national, regional and international levels to improve the economic status of the populations concerned.

SUSTAINABLE DEVELOPMENT STRATEGIES (SDS)

Increasing capacity to cope with extreme events

Capacity to cope has global, regional, national and individual perspectives. At the global level, the impact of rapid communications and the electronic media is such that it is no longer possible for any country to insulate itself against the economic and moral implications of calamities befalling mankind. This was demonstrated during the serious drought and subsequent famine which affected the Sahel during the early 1980s. The world responded belatedly, but enthusiastically, with emergency supplies of food and other life-saving amenities. A number of development projects and programs such as the United Nations Program of Action for African Economic Recovery and Development (UNPAAERD) and numerous bilateral donor projects have also been initiated. In addition, studies have been focused on the world climate and possible causes of extreme events such as droughts and floods, and there is increasing interest in the monitoring of the ecological systems bordering the Sahara in an effort to establish the long-term trends and their possible implications for the economies and social lives of the populations. It is hoped that through such initiatives the world will evolve a more useful capacity to cope with extreme events wherever they occur in the future.

Extreme climatic variations are bound to affect entire regions irrespective of national boundaries. It is therefore encouraging to see a strong concern emerging regarding the need for regional collaboration between nations in various parts of Africa. It is hoped that this spirit will lead to various degrees of regional integration and improved capacity for joint or coordinated actions in safeguarding populations from hazards of extreme climatic events. The AGRHYMET Center and DMC have an important role in the development of SDS by providing a better understanding of the climate characteristics and early warning of severe climatic variations at the regional level.

Whatever projects and programs are deemed necessary at global and regional levels, the responsibility for their implementation rests with national governments. At this level, the general statements have to be translated into policies, plans and programs with the attendant allocation of resources. It is at this level that the issues of sustainability should be rigorously addressed, since the governments will have to find the resources required and deal with any repercussions arising from the strategies adopted. The question of technology needs to be discussed in this respect. Experience has repeatedly shown that large-scale irrigation projects are expensive to install and to maintain, and may not eventually be sustainable on a long-term basis. It is therefore necessary for governments to reconsider the wisdom of such major investments and to encourage greater utilization and stepwise improvements of the time-tested traditional technologies.

Finally, sustainability of development strategies will depend on the communities affected and their perception of development itself, which may not always coincide with that of the government planners. In this respect, the future of nomadic pastoralism is an important issue for the Sudano-Sahelian region and requires more attention from ecological, economic and sociological viewpoints. Climatologically, the most important parameter in the region is rainfall, which is markedly erratic and seasonal. Nomadic pastoralism has evolved over centuries as the most sustainable, low-technology, land-based economic system. It has evolved complex land rights which ensure equitable access to the land and water resources as well as controls which enhance the regeneration of vegetation. This system has in the past provided employment through division of labor and community-based food and social security. Although these areas could

not be expected to carry large populations, population growth was not a problem for nomadic pastoralists.

It is unlikely that the region or even the world will find sustainable alternative economic activities in the foreseeable future for over 30 million nomadic pastoralists who occupy the Sudano-Sahelian region, yet their survival is now increasingly threatened by the introduction of agriculture and to a certain extent tourism through demarcation of game reserves. Nomadic pastoralism is also bound to decrease in the long run as the younger generations gain education and opt out of the system. It is therefore necessary to re-define 'development' in the context of this group of people and to orient future development strategies toward meeting their changing needs within the context of their natural ecosystems. Such strategies should exclude massive translocation of populations, change of land tenure and encouragement of sedentary agriculture as options in development.

This strategy does not provide answers to the problem of the arid areas in the long term, or those of settlement and employment of the large populations being generated in these and other parts of the region. A certain number of people can certainly be accommodated in parts of the semi-arid areas where there is a good potential for irrigation, or where rainfall is distributed in such a way that successful and economic crop production is possible at least three years in five. Others will, however, have to find accommodation and alternative livelihood systems. SDS for the Sudano-Sahelian region must therefore include policies, plans and projects designed to generate employment and alternative livelihood systems within the densely populated rural and urban areas. Rural–urban migration will then cease to be seen as a problem. Rather it will be a welcome, even if inevitable, strategy for alleviating excessive pressure on the pastoral and arable lands. An example of this strategy is the encouragement of the informal sector in a number of African countries. The informal sector activities will, however, not be sustainable unless they are integrated with the other production sectors of the economy. Such integration would improve the quality and hence the markets for goods and services generated by the informal sector.

Implications for planning and financing the development process

Change of land use

At the national level it should be recognized and accepted that most of the land in the Sudano-Sahelian zone is too dry for arable agriculture. Attempts to change the economic base to crops may look attractive on a short-term basis, but are bound to fail eventually as the yields fall and production becomes risky and uneconomical. The end result of this is severe soil degradation where such cultivated rangelands are abandoned. This has happened in some parts of the Sudan. The future of these lands is therefore strongly bound, at least in the short and medium term, to livestock production in range systems. Development plans and projects should therefore be geared primarily to the prevention of over-stocking and overpopulation through incentives rather than coercion.

Diversification

At the individual or family level, climatic variability will certainly increase uncertainties regarding the already meager sources of food and income. Capacity to cope will therefore entail development of a broader range of skills, greater mobility and adaptation to a wider range of staple foods.

However, all countries in the region do have areas of higher potential, where under good management the rainfall is adequate to grow a crop. It is these areas which will have to provide food security and sources of income and employment for the majority of the population.

Alternative livelihood systems and the problem of environmental refugees

While the immediate problems of environmental refugees can be alleviated by emergency relief and food for work programs, certain measures need to be taken in order to reduce the problem in the short term and medium term. These include:

- increase value added by processing primary products such as hides and skins, gum arabica and honey within the drylands where they are produced;
- put more effort into the development of drought-resistant crops;
- improve the infrastructure, especially those aspects affecting marketing or livestock;
- encourage inter-state trade in order to retain value added on commodities within the region;
- encourage free movement of labor;
- improve prediction of climatic trends.

The problem of getting whole communities of destitute people to settle in new environments or even to go back to their original habitats should not be underestimated. This is particularly important where such people have stayed in temporary settlements for long periods. Societal organization and values may then have changed irreversibly and original skills for coping with climatic uncertainties may already be forgotten. Long-term SDS will therefore require careful documentation of case studies on various technologies and experiences in specific areas. Global or regional data are of little value in such exercises.

It is therefore clear that planning and financing of the development programs in the region must take into account sociological, environmental and economic factors. The social factors include:

- impact of development on the cohesion within family and community systems;
- alleviation of poverty and inequalities within the population;
- interference with fundamental human rights of the individual, including freedom of movement and personal development: popular participation based on the concept of partnership.

The environmental factors include:

- impact of development on the sustainability of ecological balance;
- possible contribution to regional and global climate change;
- degradation of the soil;
- loss of biodiversity of flora and fauna;
- deterioration of water resources.

The economic measures include:

- development of rural alternative livelihood systems, including infrastructure;
- efficiency of marketing systems;
- increasing value added on local products.

Flexibility of development plans and processes

SDS must also take into account the uncertainty of time lags between various changes and their consequences. Developmental efforts should therefore be backed by a sustained research effort and a move away from ad hoc measures of drought relief toward a more systematic effort to increase resilience to drought. The latter requires improved agroclimatic information services performed by competent interdisciplinary teams of scientists.

Effects on climate change could also be alleviated through adjustments at the national level of development plans, and of the planning process itself, with greater emphasis on sustainability and a shorter response time. At present, national development plans in most countries are formulated for 5- to 10-year periods. Furthermore, formulation of the attendant programs and projects, allocation of the necessary resources and the start of implementation may be delayed for several years or even abandoned as political priorities and availability of resources change. The impact of development projects is therefore felt at least 10–15 years from the time ideas were formulated. Plans and programs which take such a long time to implement could easily be rendered irrelevant by climate change, with consequent loss of valuable time and resources. Development planning should, however, incorporate a high degree of flexibility so that adjustments can be made.

Table 13. *Mean seasonal rainfall in selected areas of Machakos district, Kenya*

	Long rains			Short rains		
	1957–71	1972–90	% change	1957–71	1972–90	% change
Kabaa	312.86	354.39	13.3	341.77	357.77	4.7
Kangundo	362.94	365.60	0.7	321.26	316.31	−1.5
Katumani	304.39	307.41	1.0	308.83	263.72	−14.4
K. Mawe	318.83[a]	290.48	−8.9	301.89	349.53	15.8
Kibwezi	263.90	238.61	−9.6	428.45	398.56	17.0

Note: [a] Period 1962–71.
Source: ODI/MRDASW/UON (1991).

Table 14. *Population density in Kenya by agro-ecological zones, 1932–69*

		Hectares/person			
Zone	Area (km²)	1932	1948	1963	1969
III	1 104	1.11	0.72	0.51	0.47
IV	2 158	2.12	1.60	1.11	0.87
V	5 069	22.70	12.90	6.38	2.93
VI	4 247	80.30	33.30	16.10	8.52
Total	12 578	5.50	3.70	2.43	184

Source: Lynam (1978).

Table 15. *Actual net population flows in Kenya to various agro-ecological zones*

Zone	1932–48	1948–63	1963–9
III	−17 250	−28 363	
IV	+11 940	+9 736	+17 170
V	+6 772	+19 973	+75 326
VI	+5 119	+7 036	+17 385
Total	0	0	+47 175

Source: Lynam (1978).

that farming techniques have evolved to support an increasing population.

Population. The historical population and intra-district migration data for Machakos in the past 60 years are shown in Tables 14–16. According to the 1979 census, the population was distributed as follows: zones II and III, 22%; zone IV, 37%; and zones V and VI, 35%. Six percent of the population was in the urban areas of the district (ODI/MRDASW/UON 1991).

The deteriorating per capita land-holding in all zones should be noted. In addition, there has been a dramatic increase in the population of the less humid zones as a result of migration from the more humid areas. Finally, there was a dramatic jump in the population of zones V and VI in the 1960s as a result of abolishing the colonial land movement structures.

Soil erosion. As early as the 1930s, the Machakos district was considered one of the most eroded areas of the country. The situation continued to deteriorate through the 1940s, but from the 1950s to date, the farming systems which emerged began to manage the cultivated land, and deterioration was

Prospects for economic, social and environmental sustainability

Future prospects for the region in terms of sustainable development will depend on the adoption of SDS which build on accumulated experience through participatory planning, and integrated development which avoids over-reliance on a few commodities, and extension of technological information to the lowest units of development. This is illustrated by the following case studies.

Experience with evolution of participatory planning and integrated development in Machakos district, Kenya

Introduction. Only about one-third of the Machakos district receives the 750 mm of rainfall that is the minimum for cropping in a bimodal regime (ODI/MRDASW/UON 1991). There are no clear data on long-term cycles, but inter-annual variability of rainfall is a dominant climatic factor. Only a small part of the district is able to receive 250 mm each season for 6 out of 10 years (Table 13). Yet it is in this system

Table 16. *Percentage population growth rates in Kenya by agro-ecological zones, 1932–69*

Zone	1932–49	1948–63	1963–9
III	2.80	2.30	1.60
IV	1.75	2.50	2.25
V	3.50	4.80	13.70
VI	5.60	5.00	11.00
Total	2.50	2.80	4.80

Source: Lynam (1978).

Table 17. *Livestock distribution in Kenya by agro-ecological zones, 1944, 1956 and 1981*

Zone	SU 1944	SU 1956	ha/SU 1944	SU 1981	ha/SU 1981
Zones II and III	49 200	79 425	2.11	12 477	6.06
%	23	22		4	
Zone IV	151 400	143 190	1.56	125 073	3.13
%	71	40		37	
Zones V and VI	13 450	133 513	3.18	196 324	4.30
%	6	37		59	
Total	214 050	356 128	1.79	333 874	4.38

Note: SU, Stock unit (livestock unit).
Source: ODI/MRDASW/UON (1991).

mainly in the communal grazing areas and lands opened for cultivation in drier areas (ODI/MRDASW/UON 1991). Soil erosion was not a problem in areas affected by tsetse flies since no agriculture could take place there, but some of these areas have been claimed for agriculture and the situation may change. Satellite imagery and studies on the ground have established definitively that the rate of erosion has slowed down where land was put under smallholder cultivation and controlled grazing. It should be noted, though, that grazing land is generally more eroded since the techniques practiced on crop land are not used on grazing land.

Crop production. There was doubt in the 1930s as to whether the district could feed itself in the 1930s. However, recent studies of the district show that it has not only produced enough food, but has evolved a complex farming system which appears sustainable. Abolition of the colonial economy which restricted some cash crops to the colonizer has led to a significant growing of such crops. The recent development of export horticultural crops has created a niche to be exploited under irrigation which was unknown in 1963. However, of more significance is the continuing intensification of food crops for subsistence and the national market (ODI/MRDASW/UON 1991).

The growing population has increased the cropped area to support the production of basic foodstuffs (mainly maize and pulses) and has diversified between crops, in terms of food and cash crops and horticultural crops (ODI/MRDASW/ UON 1991).

Livestock. As with food crops, not only has livestock production increased, but it has done so under a complex production system which has differentiated into stall feeding, tethering and range use under pressure of the shrinking land resource. Relative to 60 years ago there are more cattle in the district, but the per capita holdings have remained relatively stable (ODI/MRDASW/UON 1991).

Over the past 60 years the production areas for livestock have changed, with very little livestock in the more humid zones and more and more livestock moving into the drier zones (Table 17).

During the colonial period, the goat was the worst culprit of land degradation. It received such a bad press that very little veterinary and production research was performed on it. Yet from the data available (ODI/MRDASW/UON 1991), it is clear that the goat is an important component of the livestock farming system of the people, and the ratio of goats to cattle has essentially remained stable over a very long period. This has major implications for Sudano-Sahelian development strategies which ignore small stock. Small stock are preferred for their immediate conversion in exchanges and in meat production. They are also ecologically more viable in the less humid areas. Their production suggests a non-ranch strategy which maximizes production of grass, herbs and shrubs, and even trees to produce shoat (sheep and goat) fodder.

The livestock marketing data show clear adjustment to drought pressure. This is in direct contrast to many studies from the Sahelian region which suggest that the primordial attachment to livestock is still operative and producers wait until droughts wipe them out (ODI/MRDASW/UON 1991).

Land use and conservation. The increase in production has been accomplished under a complex land use system in which management of vegetation for production and conservation reasons is practiced so systematically that the fears of degradation expressed in the 1930s have been reversed (ODI/ MRDASW/UON 1991). The premier soil and water conservation technique has been terracing of crop land, which was introduced in the 1950s. This is usually done by work groups, as project-related tools-for-work or food-for-work, family labor or hired labor. Contour bunds were introduced in the 1930s, but farmers know they are not as effective as terracing. Other techniques are reforestation or management of natural biomass in grazing areas which do not get as much conservation attention as cropland.

Conservation of the natural vegetation has improved in the range areas in spite of fears in the 1930s that it would be abandoned. These areas, which are essentially in zones V and VI, produce most of the livestock in the district (ODI/MRDASW/UON 1991). It should be emphasized that the production system does not assume the classical ranching system so preferred by development specialists. The emergent system is essentially an indigenous system with mixed livestock which does not strive to maximize grass at the expense of herb and shrub species.

Farmers' organizations. To cope with the evolving system, farmers have modified traditional organizational systems to accomplish new and modern tasks. One example is a farm in zone V which has gone from traditional communal grazing in the early 1900s to world-standard dairy farming by 1990. Through modification of traditional institutions, Machakos farmers have organized cooperatives extensively. In 1987, 138 452 farmers were members of 54 primary cooperatives organized into a District Cooperative Union. Given the average family size of eight, benefits from this movement were reaching approximately 90% of the district population. The 54 cooperatives had a subscribed share capital of Ksh. 18m (US$ 1m) and a turnover of Ksh. 465 651 776 (US$ 25 869 543) (Mutiso 1988).

The Machakos lesson. The Machakos study suggests that society can, within limits, evolve farming systems which produce efficiently, even in drier regions. This suggests that a climatic change which leads to less humid conditions need not be a disaster if SDS are in place. This case study shows further that semi-arid areas can be self-sufficient in food and contribute to national development given sufficient political will, appropriate developmental strategies and programs which originate and are implemented at community levels, and a reasonable level of available resources to initiate development.

Adaptation to drought: experiences in Danbatta district of Kano state in Nigeria
It is interesting to compare experience in the Machakos district with that of the Danbatta district in Nigeria as described by Mortimore (1989). Unlike Machakos, Danbatta started off in relative prosperity. The onset of the Sahelian drought, however, found the district unprepared for an extended emergency, resulting in famine, loss of livelihood and reduction of the ecosystem carrying capacity. Traditional coping systems were tested to their limit, but they failed to produce solutions which could sustain the population through prolonged drought. The population has there-

fore been forced to evolve new strategies for coping, but their impact is unlikely to be felt beyond mere survival.

The environment and traditional practices. The Danbatta district occupies an area in the dry zone of northern Nigeria which continues into adjacent areas of the Niger Republic. The southern boundary experiences 1000 mm rainfall in a 40-day growing season, but rainfall decreases to 500 mm in a 60-day season in the north. The region experienced relatively stable conditions, culminating during the 1960s in the highest levels of agricultural productivity based on export crops, especially groundnuts. This stability was taken for granted and no measures were taken to anticipate a possible dramatic change of fortune. The ecological stability was maintained within acceptable limits by use of manure, rotational bush fallows and by migration to less densely populated areas. Domestic energy requirements were met from wood fuels derived by cutting living trees in the bush and on farmland. These resources were supplemented with crop residue and charcoal. Nomadic cattle herds were managed in a predominantly north–south 'transhumance' system which supplied most of the meat consumed in the country. Only a few cattle remained in the same location throughout the year, depending on availability of water and fodder, and most animals were sold when fully grown.

Climatic change and impacts. Rainfall amounts in the region began to fall around 1968, and by the end of 1972 a vast area of West Africa was in the grasp of drought and crop failures reached 50–90%. Most households in the Danbatta district consumed all their subsistence grain, and livestock losses were estimated at 4400 cattle, 5000 goats, 500 donkeys and most of the poultry. Groundwater levels fell, leaving dry wells and necessitating repeated deepening of the wells. Food prices increased by over 40% and grain supplied by the government made little impact due to the large population. Hunger and sickness increased.

Societal responses. Societal response took many forms, mainly derived from social relationships and improved mobility. Alternative foods were sought in the market, and employment opportunities, selling of manure and selling of personal property decreased. There was, however, an increase in selling of livestock, wood and mats, in dry season migration, and in selling of land and begging. The population tried to extend the area cultivated to compensate for low yields, but these efforts were limited by shortage of seed. There was a reduction in fertilizer use, changes in crop mixtures and a shift away from guinea corn, groundnuts and cowpeas, which were considered risky investments.

In general, the population seemed to accept their fate and to adapt to poverty by finding alternative opportunities in petty trade (e.g. in fuel wood), liquidation of assets, mobilization of traditional social support systems and migration to Kano city.

Sustainability. It appears that although the traditional coping system based on subsistence seemed to break down during the first few years of drought, it was resilient and was already showing signs of recovery by 1974. Time will tell the extent to which these strategies will be sustainable, especially as the inevitable changes in social and economic needs and aspirations of the population take place. The test for sustainability cannot be done from global data alone. Detailed case studies to assist governments to evolve adequate policies and strategies are needed.

Integration of agroclimatic information in agricultural production: the Mali experience

Introduction. The climate of the Sudano-Sahelian region is characterized by marked seasonality, spatial and temporal variability of rainfall, and high evapotranspiration rates. Droughts are therefore part of, rather than anomalies in, the Malian climate, and a drought is experienced twice every 3–5 years (Konare 1989). The mainstay of the Malian economy is agriculture, and 85% of the cereal crop is produced under rainfed conditions. Agroclimatic information should therefore be able to play a crucial role in minimizing production risks which result in losses of seed and crop yield. A great deal of agroclimatic data has been collected and research has demonstrated the potential benefits of such data in agricultural production. This knowledge, however, has not had a significant impact on the traditional farming practices and an innovative approach was developed to make meteorological information available to farmers.

The Pilot Project in Agrometeorology. The traditional knowledge of the Sahelian farmer has proved inadequate to cope with current disturbances in the traditional crop calendar as a result of climatic variations. An experimental project entitled the Pilot Project in Agrometeorology was therefore designed and has operated in Mali since 1982 (Konare 1989). The project is located southwest of Bamako in a sector of Bancoumana which is part of the agricultural extension organization – Opération Haute Vallée (OHV). The project is managed by a multidisciplinary team of agronomists, meteorologists, crop protection and functional literacy specialists. The team was responsible for analysis of agrometeorological and other relevant information and for deciding on the appropriate advice and warning to be given to farmers. This information forms part of the technological package designed to improve agronomic operations such as time of planting, ploughing, weeding and crop protection. It

is derived from rainfall statistics, meteorological forecasts, 10-day estimates of water balance and crop water requirements. A group of farmers identified by the team participated in the project by setting up two plots in their farms. The participating farmer was required to manage the two plots in the same way except that agrometeorological advice from the team would be applied to the 'test' plot. The farmer and the team are then able to compare the yields from the two plots at the end of the season. From the results, the farmer is free to choose whether to adopt the recommended practices on the rest of the farm while the team is able to identify aspects which need improvement.

Outcome. Feedback obtained during field trips to assess the project indicated that farmers were convinced of the value of the agroclimatic information they applied. Data analysis showed that yield increases of 26% for sorghum and 46% for millet were achieved in the test plots managed according to the team's advice (Traore and Konare 1989). The experiment has succeeded in convincing the team of the value of an integrated approach in extending new technologies to farmers, and the potential for application of agroclimatic information for sustainable agricultural production. The farmers have also become increasingly confident, and they adopt the recommended practices as indicated by the large increase in the number participating. The results from this experiment have other implications as follows:

- the benefits of application of agrometeorological information increase as climate variability increases from semi-humid through semi-arid to arid regions;
- yield increases reduce the necessity for farmers to expand cultivated land with the attendant risks of land degradation;
- use of meteorological information enables optimum use of agro-chemicals thus reducing their harmful effects on the environment;
- these experiments help in the establishment of a data base for research on the environment.

CONCLUSIONS AND RECOMMENDATIONS

Regional prospects for sustainable development without climate change

The Sudano-Sahelian region is not homogeneous environmentally, socially, economically or politically. The region is, however, characterized by climate limitations and variability (mainly low and erratic rainfall) which recurrently threaten the very existence of human and livestock populations. In spite of these handicaps, the region has sustained a growing population for thousands of years, even though at a relatively low level of socioeconomic development. The region is, however, now considered to be at high risk as a result of population increase, and political and socioeconomic developments which have disrupted traditional coping strategies without offering viable alternatives. The result has been varying degrees of degradation of the environment, reduced land carrying capacity and increasing frequency of human suffering.

Sustainable development in the region requires a more complete understanding of climate variability, as well as recognition and strengthening of the traditional complementarity of the various ecological zones. The new phenomenon of urbanization will also have to be accepted and incorporated within national development policies, plans, strategies and programs. This chapter argues that prospects for sustainable development under prevailing climatic conditions are good, provided development policies put less emphasis on 'change' in favor of 'progressive improvement' in those strategies and technologies which have enabled the populations to cope in the past.

(1) The key to environmental conservation in the arid and even parts of semi-arid areas which are most prone to climate variation is maintenance of vegetation cover. The mixtures of grass and woody species which have evolved in these areas are adapted to seasonal variations in rainfall and bush fires, which are used in traditional pasture management by nomadic pastoralists. Changes of land use from nomadic pastoralism to grazing blocks, group ranching and even arable agriculture, are bound to fail with resultant land degradation. The populations in these ecozones should therefore be assisted to cope with their natural environment through:

- health and educational services which do not require sedentarization;
- socioeconomic and livestock marketing opportunities which encourage pastoralists to keep only the productive animals and a better balance between cattle, camels, sheep and goats;
- improvements in traditional production systems in agriculture and livestock production;
- encouragement and facilitation of migration of excess population to better opportunities within or outside the country, by providing training in skills which are in demand elsewhere. Such migrants could provide valuable financial support to their families through remittances, which is already happening.

(2) Sub-humid and some parts of semi-arid areas, while still fragile, offer better opportunities for increased food production and agro-based industries. It is in these areas that investments in soil and water conservation can be translated into more intensive agriculture and livestock production, providing internal food security and employment opportuni-

ties for the bulk of the population – as has been demonstrated in the case study on the Machakos district of Kenya. The main challenge is to ensure sustainability of development efforts. This can be achieved through:

- ensuring that development programs are community based in both origin and implementation;
- ensuring that development programs build on, rather than supersede, the traditional systems of land tenure, division of labor, food preferences, etc.;
- diversification of crops and economic activities.

(3) A major potential threat to the environment and sustainability in these regions is not climatic variability *per se*, but the combined effects of an increasing population, the inadequacy of traditional technologies, a low level of economic development and lack of alternative employment opportunities, which lead to an overutilization of land resources. Under these conditions, soil fertility is bound to deteriorate as rest periods in shifting cultivation become shorter, and crop rotations become more and more difficult. The result is reduced carrying capacity as yields decrease below subsistence levels. Strategies to cope with this situation include:

- in the short term, establishment of labor-intensive industries and small-scale irrigation schemes;
- in the long-term, reducing the rate of population growth through education and appropriate economic incentives;
- promotion of alternative livelihood systems which would create employment opportunities in both rural and urban areas in order to reduce pressure on the productive land resources.

Potential for climate change to undermine sustainable development

If the climate does change dramatically in the future, and depending on the speed and direction in which such changes occur, it is possible that sustainability of a number of developmental strategies and programs could be undermined with considerable losses to both national economies and individuals. If, for example, the region dries out on a long-term basis, the present ecozones VI and VII would become uninhabitable deserts. Some of the consequences would be:

- massive migration of the pastoralists and their herds to the remaining agricultural lands;
- all plans based on complementarity between ecozones would collapse, and the resulting land pressure could precipitate major social and economic hardships;
- the irrigation schemes would also go out of production for lack of water, undermining strategies for food security and creation of employment.

The effects of a major increase in rainfall could also undermine sustainable development by:

● destroying dams and flooding agricultural land in the fertile river valleys;
● disrupting surface transport and communication systems.

Rapid changes in climate would also undermine sustainable development by forcing frequent changes in development plans, including strategies and programs.

Adjustments to enhance sustainability

Although there are speculations that major climatic changes could occur in the region as a result of global warming, there is no clear consensus on whether the rainfall regime will improve or deteriorate. It may, therefore, be more appropriate to assume that both long-term and short-term variability in rainfall will continue, and that the future of the populations will depend on measures which can be put in place immediately to arrest and possibly reverse environmental degradation in a sustainable manner. This is not impossible to achieve, but much more emphasis will have to be put on the issue of sustainability. As observed at the FAO/Netherlands Conference on Agriculture and Environment (FAO 1991), the traditional extensive production system with a low capital/labor ratio appears to be the most rational and economically viable in a high-risk agricultural economy such as the Sudano-Sahelian region. Sustainability will be enhanced by:

● intensification of climatic and ecological monitoring systems at national, regional and global levels in order to provide early warning of dangerous trends;
● establishment at the national level of plans and strategies to deal with emergencies caused by floods and intensive droughts;
● improvement of information on climate, and ecological changes, and integration of the roles of information, technology and public interventions;
● diagnostic studies and research to investigate global aspects of droughts that can facilitate their prediction;
● provision of meteorological advisory and information services to end users in order to increase agricultural production;
● better water management, including water harvesting and development of small-scale irrigation schemes.

In national planning and financing processes, adjustments are also required to shorten the lead time in both planning and implementation of projects. Short-term plans will also have to concentrate on optimal utilization of resources in order to establish economic buffering systems, while long-term plans should emphasize the role of research and development of a wide range of technological skills. A key issue in all these strategies is the source of energy. The region has large resources of solar and geothermal energy which are still untapped. Programs to develop the use of these resources could facilitate industrial development and reduce the rate of deforestation and land degradation.

REFERENCES

Bolin, B. 1989. Statement at Ministerial Conference on Atmospheric Pollution and Climate Change. Noordwijk.
Brown, L. H. 1963. The development of the semi-arid areas of Kenya. Unpublished paper.(Mimeographed.)
David, R. and Myers, M. 1991. Nature's nurturers hit by male exodus. *Panascope* **26**:17.
FAO 1991. Sustainable agriculture and rural development in sub-Saharan Africa: Regional Document 1: the Den Bosch Declaration and agenda for action on sustainable agriculture and rural development. FAO/Netherlands Conference on Agriculture and Environment.
FAO/UNFPA/IIASA 1984. Technical report FPA/INT/513.
GOK 1979. *Arid and Semi-arid Land Development in Kenya: The Framework for Implementation, Program Planning and Evaluation.* Government of Kenya.
GOK 1983. *District Focus for Rural Development.* Government of Kenya.
GOK 1989. *Agriculture and Livestock Data. MOPND Long Range Planning.* Govrenment of Kenya.
GOK 1990. ASAL Development Policy Paper Draft 3. Government of Kenya.
Hammond, A. L. and Paden, M. E. (eds.) 1990. *World Resources 1990–91.* New York: Oxford University Press.
Hare, F. K. 1985. *Climate Variations, Drought and Desertification.* WMO no. 653. World Meteorological Organization.
Hellden, U. 1984. *Drought Impact Monitoring.* Lund University Naturgeografiska Institut, Sweden.
Hulme, N. 1989. The changing rainfall resources of Sudan. *Transactions of the Institute of Geography, New Series* **15**:21–34.
Jaetzold, R. and Schmidt, H. 1983. *Farm Management Handbook of Kenya: Natural Conditions and Farm Management Information.* Nairobi: Ministry of Agriculture.
Konare, K. 1989. Applied meteorology in the Sahel: the Malian experience. *Weather* **44**(2):64–70.
Lamprey, H. F. l988. Report on the desert encroachment reconnaissance in Northern Sudan. *UNESCO/UNEP Desertification Control Bulletin* **17**:1.
Lynam, J. K. 1978. An analysis of population growth, technical change and risk in peasant, semi-arid farming systems: a case study of Machakos District Kenya. PhD Thesis, Stanford University.
Mortimore, M. 1989. *Adapting to Drought: Farmers, Famines and Desertification in West Africa.*
Mutiso, G.-C. M. 1991. *ASAL Institutions Files.* Muticon Ltd.
Mutiso, G.-C. M. and Mutiso, S. M. 1991. *Kambiti Farm: Water in Capitalising ASALs.* Muticon Ltd.
Mutiso, G.-C. M. *ASAL Institutions Files.* Muticon Ltd.
Nicholson, S. E. 1989. Long-term changes in African rainfall. *Weather* **44**(2):46–56.
Nicholson, S. E., Jeeyoung, K. and Hoopingarner, J. 1988. *Atlas of African Rainfall and its Inter-annual Variability.* Florida State University, USA: Department of Meteorology.
ODI/MRDASW/UON. 1991. Environmental change and dryland management in Machakos District, Kenya, 1930–1990. Draft Working Papers, September 1991: 1. Conservation profile, F. N. Gichuki; 2. Crop production, S. G. Mbogoh; 3. Land use profile, R. S. Rostom, M. Mortimore and J. Yego; 4. Livestock production, C. Ackello-Ogutu; 5. Natural vegetation, K. O. Farah; 6. Population profile, M. Tiffen; 7. Rainfall, S. K. Nutiso; 8. Soil erosion, D. Thomas; 9. Soil fertility, J. Mbuvi.
Parry, M. L., Carter, T. R. and Konijn, N. T. (eds.) 1988. *The Impact of Climatic Variations on Agriculture,* vol. 2, *Assessments in Semi-Arid regions.* UNEP/IIASA.
Republic of Senegal 1991. *Vers un Développement Durable.* National

Report for UNCED.

Tegart, W. J. McG., Sheldon, G. W. and Griffiths, D. C. (eds.) 1990. *Climate Change: The IPCC Impacts Assessment*. WMO/UNEP.

Traore, K. and Konare, M. 1989. Operational agroclimatic assistance to agriculture in sub-Saharan Africa: a case study. In *Climate and Food Security*, pp. 193–202. Manila: International Rice Research Institute.

UNDP 1991. *Human Development Report*.

UNEP 1991. *Global Assessment of Desertification*. UNEP.

UNFPA 1991. *World Population Prospects. UN Population Studies 1990*. New York: United Nations.

UNSO 1991. *Drought and Desertification in the Context of UNPAAERD 1986–1990*. UNSO.

UNSO 1992. *Assessment of the Drought and Desertification in the Sudano-Sahelian Region 1985–1991*. UNSO.

WMO/UNEP 1990. *1990 Climate Change: The IPCC Response Strategies*. World Bank 1990. *World Development Report*.

WMO/UNEP 1991. *The Global Climate System: Climate System Monitoring*, p. 21.

World Bank 1989. *World Development Report*.

World Bank 1990. *World Development Report*.

4: Climate Change and Sustainable Development in China's Semi-arid Regions

ZONG-CI ZHAO

INTRODUCTION

This chapter sketches the historical and current characteristics of West China's semi-arid lands, designated herein as IMJSHNG, an acronym for the five provinces affected. Data on the frequency of droughts and floods go back as far as the 1400s and as early as 180 BCE (Before the Common Era) for some major climatic events. The temperature data for this region seems to show a trend toward warming of between 0.5 and 1.8 °C. The chapter also discusses some major climate- and population-related problems facing the IMJSHNG region in recent years. General circulation models (GCMs) have been used to simulate future climate change for these regions, and while models clearly project warming, as is the case with other semi-arid regions of the world, generalized predictions are difficult to make. Because approximately 21.7% of the total area of China consists of semi-arid land, studies of the ecosystem, the environment, climate changes, and the society of this area are vital. Sustainable development and policy strategies relating to this area are also key issues for study.

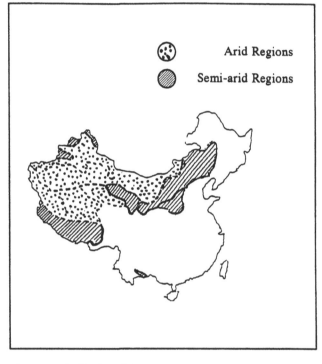

Figure 1: Distribution of the arid and semi-arid regions in China. (Note: Some Chinese islands are not shown.)

BASIC CHARACTERISTICS

Location

The semi-arid regions of China are generally located in the west and northwest of the country. About 70% of the semi-arid regions consist of hills, plateaus above 1000 meters (m), and mountains. Approximately a quarter of these semi-arid lands are located in Inner Mongolia autonomous region, in the Shanxi, Shaanxi, Ningxia and Gansu Provinces or autonomous regions and in parts of the Qinghai, Xinjiang, Xizang and Yunnan Provinces or autonomous regions (Fig. 1). Located between the arid and semi-humid regions of China,

these semi-arid lands are subject to strong winds, blowing dust, sandstorms and shifting dunes.

Because of the economic importance of their agriculture and animal husbandry, the semi-arid regions have long been the focus of Chinese scientists and policy makers (Song Shi-da *et al.* 1980; Lu Yu-rong and Gao Guo-dong 1985; Zhao Song-san *et al.* 1985; Geng Kuan-hong 1986; Lin Zhi-guang and Zhang Jia-cheng 1986; Chinese Natural Resource Research Committee *et al.* 1988; Li Jiang-feng *et al.* 1990; Zhao Zong-ci and Zang Xuejun 1991).

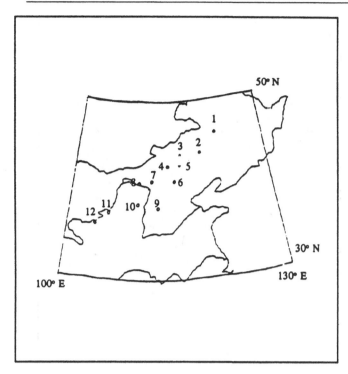

Figure 2: Location of the 12 stations in the semi-arid regions. 1, Wulanhoute; 2, Lindong; 3, Xilinhoute; 4, Wenduermiao; 5, Dulun; 6, Zhangjiakou; 7, Huhehoute; 8, Baotou; 9, Taiyuan; 11, Yulin; 11, Zhongning; 12, Lanzhou.

Climatic characteristics

The climate in most parts of Inner Mongolia, Shanxi, Shaanxi, Ningxia and Gansu Provinces is very similar (abbreviated to IMJSHNG). This region has the climatic characteristics of the monsoon, with cold, dry winters and warm humid summers. The annual mean temperature ranges from 5 to 10 °C. The annual total precipitation ranges from 250 to 500 millimeters (mm), and the annual total evaporation rate ranges from 250 to 400 mm. The months of July, August and September are characterized by heavy rain. Rainfall is scant the rest of the year.

Twelve meteorological stations in the IMJSHNG area have been chosen for study here. The locations of these are shown in Fig. 2.

Comparisons of various climatic characteristics of the semi-arid regions of China (IMJSHNG) with other regions of the country are presented in Figs. 3 and 4. These comparisons include statistics on: (a) annual total precipitation, (b) annual total evaporation, (c) annual runoff and (d) desert areas in the region.

Fig. 4(a)–(d) are regional maps showing yearly statistics on (a) the number of days per year the area is subject to winds in excess of 17 meters per second (m/s), (b) the annual mean wind speed, (c) the number of days per year during which

sandstorms occur and (d) the number of days of flowing sand. As noted in the figures, there is less annual precipitation, evaporation and runoff in the semi-arid regions than in the rest of the country: precipitation is 250–500 mm, evaporation 200–400 mm and runoff 25–100 mm.

In the IMJSHNG area the number of days of mean strong wind (wind speed more than 17 m/s) ranges between 25 and 75 per year. In other parts of China the statistics show strong wind on from 5 to 20 days per year. Again, in IMJSHNG the annual mean wind speed is 2.5–5 m/s while in most other parts of China it is 1–2 m/s. There are deserts in this area.

The number of days per year that sandstorms occur (with visibility less than 1000 m) in IMJSHNG varies from 5 to 20. Only the arid regions exceed these statistics. By comparison, in most of China the annual mean occurrence of sandstorms varies between 0 and 5 days.

In IMJSHNG the average number of days per year of flowing sand (with visibility between 1 and 10 kilometers) varies from 15 to 60. By comparison, in most of China there is flowing sand on fewer than 10 days per year.

Population and economics in the Northwestern semi-arid regions

The main semi-arid region of China is located in parts of Shanxi, Shaanxi, Ningxia and Gansu (JSHNG). This area lies between Northwest China and the central part of China. It covers approximately 882 300 square kilometers, which comprises roughly 9% of the total area of China.

The area contains approximately 8% of the total population of China, or approximately 83 570 000 people. Between 1955 and 1970 the population of this semi-arid region increased very quickly due to immigration from East China to West China.

Table 1 illustrates the total population, the urban population, the mean natural population growth rate and the mean income per person for Shanxi, Shaanxi, Inner Mongolia and Gansu. Comparisons of this region with the whole of China are shown in parentheses in the table, which illustrates the population increase between the 1950s and the 1970s. The mean income per person in this area was lower than the average income of the rest of China.

Table 2 illustrates the mean natural increasing rate and density of population in five cities of this region (Datong, Huhehaute, Xining, Baotou and Linfen). The mean natural increasing rate in the cities during the 1950s and 1960s was a direct result of immigration from East China to West China. The mean density of population in the same five cities also greatly increased during this period.

A total of 158 200 000 mu of JSHNG is cultivated crop land (mu is a Chinese unit of area; 1 mu equals 1/15 hectare).

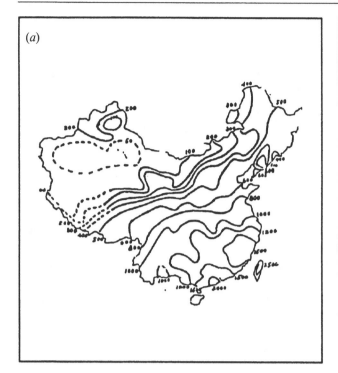

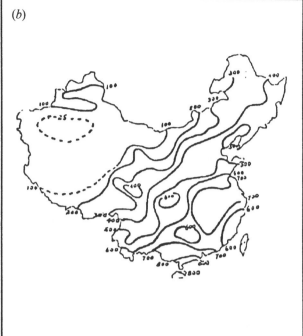

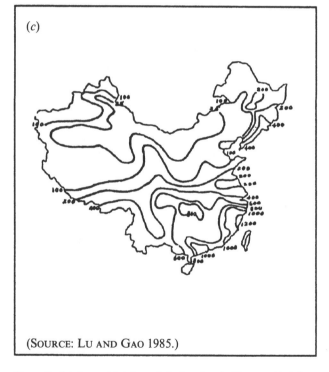

(SOURCE: LU AND GAO 1985.)

Figure 3: (*a*) Annual total precipitation (mm), (*b*) annual total evaporation (mm), (*c*) annual runoff (mm) (*d*) desert region. (Note: Some Chinese islands are not shown.)

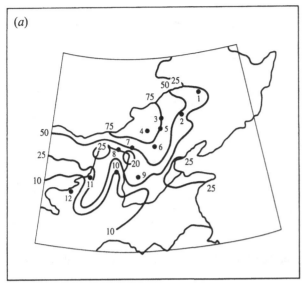

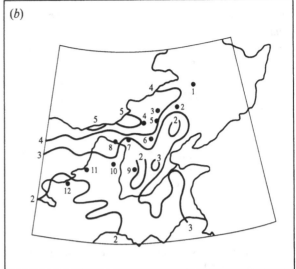

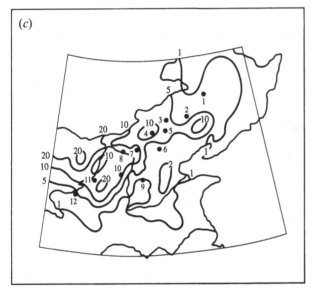

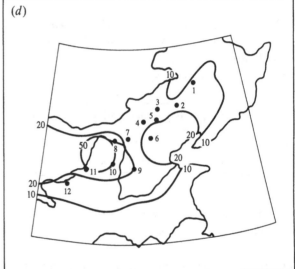

Figure 4: (a) Number of days annually of strong wind (> 17 m/s), (b) annual mean wind speed (m/s), (c) number of days annually of sandstorms and (d) number of days annually of flowing sand.

This is approximately 9.5% of the total land area of China. Crop production (23 320 000 tonnes) accounts for approximately 6% of the total production in China. Mean crop production is about 279 kilogram (kg) per person, which translates to approximately 294 yuan per person. In comparison, industrial work yields a mean value of 901 yuan per person. Production values per person in JSHNG compare unfavorably with the rest of China.

Natural resources in the Northwest semi-arid region

Major factors that influence life in the Northwest semi-arid region of China include agriculture, animal husbandry, indigenous animals, water and mineral resources.

Agriculture. Relatively small areas of IMJSHNG are used for agricultural purposes. The primary crops are wheat, corn and Chinese sorghum. Some industrial crops are cultivated as well, including cotton, peanuts, soybeans, tobacco and sugar-beets. Shifting cultivation is also practiced.

Animal husbandry. Poultry production plays an important role in IMJSHNG. In addition, animals such as cattle, goats, sheep and horses graze the abundant grasslands in Inner Mongolia and Qinghai. There are few woodlands in this region. Fig. 5 shows the distribution of vegetation in IMJSHNG.

Table 1. *The total population, urban population and mean natural population growth rate in Shanxi, Shaanxi, Inner Mongolia and Gansu*

Years	Total population (10^4)	Urban population (10^4)	Growth rate (%)	Income per person (yuan[a])
1949	4174 (7.7%)	— —	—	
1955	5223 (8.5%)	686 (8.3%)	2.15 (2.03)	133 (129)
1965	6657 (9.2%)	1011 (7.8%)	2.72 (2.84)	183 (194)
1975	8574 (9.3%)	1365 (8.5%)	1.49 (1.57)	239 (273)
1985	9686 (9.2%)	2253 (5.9%)	1.14 (1.12)	559 (672)
1989	10265 (9.2%)	2614 (4.6%)	1.50 (1.43)	935 (1189)

Notes:

The figures in parentheses are the percentages this region represents of the whole of China (in second and third columns) and those values in all of China (in the fourth and fifth columns).

[a] Yuan: Chinese unit of money (exchange rate in 1989: 1 US$ = 3.7 yuan).

Table 2. *Mean natural increasing rate (MNIR) and density (D) of population of five cities in the semi-arid region (Datong, Huhehaute, Xining, Baotou and Linfen)*

	Year					
	1949	1957	1965	1978	1985	1988
MNIR (%)	1.25	2.99	2.54	1.03	0.87	1.12
D (persons/km²)	104	317	386	600	689	733

Indigenous animals. Native wildlife such as musk deer, marten and Mongolian gazelle is plentiful.

Water resources. Water resources are a very important issue in North China, especially in the arid and semi-arid regions. Water resources include rainfall, snowmelt, ground-water, rivers and lakes. The second largest river in China, the Yellow River, flows through the region. In addition there are both fresh-water and salt-water lakes, as well as smaller rivers. Fig. 6 shows the major lakes and rivers in the region. Table 3 shows the deposits and possible development of water in the semi-arid areas (data from Zhang Chuen-yuan *et al.* 1983; Yang De-qin *et al.* 1989).

Mineral resources. There is an abundance of coal in the semi-arid regions, especially in Shanxi province. Iron, petroleum and nonferrous metals can also be found.

Difficulties of development in the semi-arid regions

Significant problems in the semi-arid regions of China include such factors as desertification, desiccation, strong

Table 3. *Deposits and possible development of water in the semi-arid regions*

Provinces or autonomous regions	Deposits of water 10^8 cubic meter/year	(%)	Possible development of water 10^8 cubic meter/year	(%)
Shanxi	448.0	(0.8)	106.98	(0.6)
Inner Mongolia	435.9	(0.7)	83.50	(0.4)
Shaanxi	1116.8	(1.9)	217.04	(1.1)
Gansu	1249.5	(2.1)	424.44	(2.2)
Qinghai	1886.6	(3.2)	772.08	(4.0)
Ningxia	181.6	(0.3)	31.62	(0.2)

Note: The figures in parentheses are the percentages each region represents of the whole of China.

winds, sandstorms, inadequate water supplies, salinization of cultivated lands, water loss and soil erosion.

VARIATIONS OF CLIMATE, ENVIRONMENT AND SOCIETY

Climate changes

Temperature

Over the last 100 years in China temperature data have been measured and recorded. Table 4 shows the frequency by decade of cold and warm years over this period in the semi-arid (Northwest China) and semi-arid/semi-humid (Huabei) regions (data from National Meteorological Center 1985).

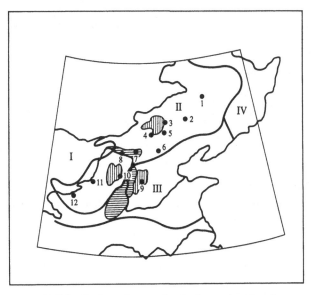

Figure 5: Distribution of vegetation in the semi-arid regions.

I   Temperate desert
II   Temperate steppe
III Warm temperate zone
     and deciduous forest
IV   Temperate mixed conifer and
     deciduous forest zone

Temperate grassland
and sand (conifer)

Crops

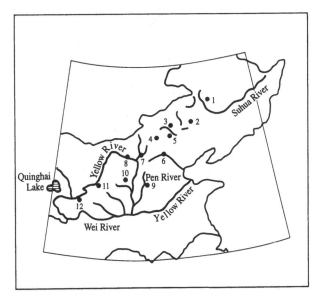

Figure 6: The major lakes and rivers in the semi-arid regions.

The cold years were defined as 0.5–1.5 °C colder and the warm years as 0.5–1.5 °C warmer than the mean climate. It can be seen from Table 4 that during the decades from 1920 to 1950 the Huabei region and Northwest China experienced a warm period, while between 1950 and 1970 the same regions experienced a cold period. However, since 1980 the temperature in these regions has been rising.

The linear trends of annual and seasonal temperature in the semi-arid regions for the last 40 years have been measured, using methods similar to those of Zhao (1991). These calculations show an increase in temperature over the last 40 years in IMJSHNG, most noticeably in winter and spring. Calculations from the 12 stations in IMJSHNG show

a linear trend of the annual mean temperatures from 0.5 to 1.8 °C. Fig. 7(a) shows the linear trends of annual mean temperature in the semi-arid region for the last 40 years.

Precipitation
Meteorological records regarding floods and droughts have been maintained in China for the last 500 years (see Academy of Meteorological Science, 1985; Shao-wu Wang and Zhao Zong-ci, 1981). The frequency of floods and droughts over this period measured in 50-year increments, is compared in Table 5, which shows the results from several stations in IMJSHNG, such as Taiyuan, Yulin, Huhehaute and Lanzhou (stations 7, 9, 10 and 12: see Fig. 2). Except for Huhehaute, these stations have maintained records for 500 years. At Huhehaute data were available only for 1720 through 1989. Several major characteristics in IMJSHNG over the last 500 years are noted in Table 5:

(1) Especially in Yulin and Lanzhou, the frequency of drought was noticeably higher than that of floods.
(2) During the years 1470–1519, 1570–1619 and 1770–1819 the climate in IMJSHNG was obviously dry, but from 1720 to 1769 the frequency of floods was higher than that of drought.
(3) Since 1970 the region has been getting dryer (see Table 5).

The linear trends of annual and seasonal precipitation in the semi-arid regions for the last 40 years have also been calculated. These show that over this period it has been getting drier in IMJSHNG, especially during the summers. The calculations from the 12 stations in IMJSHNG show that the linear trend of annual precipitation for the last 40 years is 30–70 mm (Fig. 7b).

Table 4. *Frequency (F) of the decadal cold and warm climate in the semi-arid region (Northwest China) and the semi-arid/ semi-humid region (Huabei) for the last 100 years*

Regions	Characters	1911–20	1921–30	1931–40	1941–50	1951–60	1961–70	1971–80	1981–90
Huabei	F of warm	3	6	5	8	0	1	2	4
	F of cold	5	0	1	2	4	3	5	2
Northwest	F of warm	0	3	3	8	1	0	0	5
	F of cold	3	1	1	0	3	6	3	1
Summary		Cold	Warm	Warm	Warm	Cold	Cold	Cold	Warm

Notes: Five stations were chosen in Huabei: Beijing, Taiyuan (9 in Fig. 2), Jinan, Zhengzhou and Xuzhou. Five stations were chosen in Northwest China: Yanan, Xian, Lanzhou (12 in Fig. 2), Xining and Jiouquan.

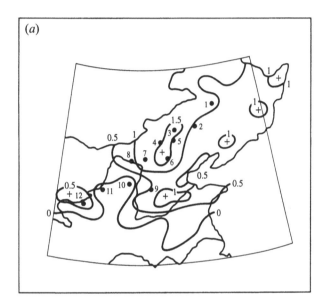
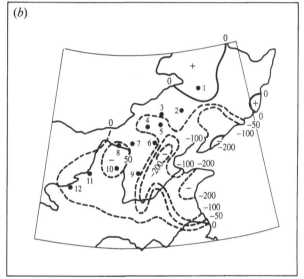

Figure 7: Linear trends 1951–89 for (*a*) temperature (°C) and (*b*) precipitation (mm).

Relationships between temperature and precipitation

The relationships between the annual and seasonal temperature and precipitation in IMJSHNG for the last 40 years have also been calculated. A negative correlation between the annual and summer temperature, and precipitation in the semi-arid regions has been noted, especially in the summer. (See Fig. 8, which shows the distribution of the significant correlation coefficients in summer.) The figures indicate that the warm climate corresponds to dry conditions in IMJSHNG in the summer. As mentioned above, from 1951 to 1990 IMJSHNG experienced warm dry conditions.

Wind

Fig. 9 shows the annual mean wind speed and the number of days of strong wind (> 17 m/s) during the last 40 years in IMJSHNG (data from National Meteorological Center,

1985). This illustrates that both the annual mean wind speed and the number of days of strong wind per year decreased over the period between the 1950s and the 1970s in most of IMJSHNG. Wind speed decreased by about 0.2–2 m/s and the number of days of strong wind decreased by approximately 5–35 during the same period. The noticeable decrease might be explained by natural climate changes, the building of shelter belts, and urbanization in this area.

Climate disasters

Statistics obtained from historical data in China (Academy of Meteorological Science, 1985) indicate that significant and serious disasters occurred in IMJSHNG and Quinghai Province 24 times during the period from 180 B.C.E. to 1949.

Table 5. *Frequency (F) of floods and droughts per 50 years in the semi-arid region of China over the last 500 years*

Stations	Characters	1470–1519	1520–1569	1570–1619	1620–1669	1670–1719	1720–1769	1770–1819	1820–1869	1870–1919	1920–1969	1970–1989	Total
Taiyuan	F of floods	8	18	12	19	19	21	18	25	21	16	9	177
	F of droughts	29	20	22	21	16	14	19	17	13	23	8	194
Yulin	F of floods	7	10	9	9	10	16	9	14	16	17	3	117
	F of droughts	31	21	21	21	12	12	24	17	23	21	10	212
Huhehaute[a]	F of floods						5	10	20	17	19	5	71
	F of droughts						14	22	18	19	15	9	88
Lanzhou	F of floods	3	8	7	8	4	14	8	11	6	14	7	83
	F of droughts	14	9	15	13	20	12	13	17	11	17	11	141
Summary		Dry		Dry			Wet	Dry				Dry	

Note: [a] Not enough data are available for Huhehaute before 1720.

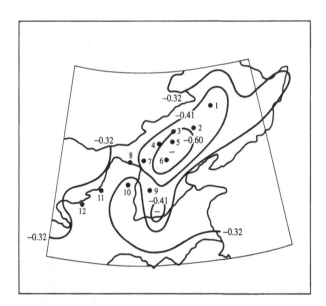

Figure 8: Correlation coefficients between temperature and precipitation in summer.

Droughts occurred 15 times, floods eight times, famine three times and epidemic disease once during this period.

Changes in population and urbanization

From 1949 to 1988 the population of IMJSHNG increased by a factor of 2–3.5. Table 6 shows the population increase in the 12 selected stations over this period. Immigration from East China to West China and from the rural areas to the city, as well as the high birth rate, accounts for the increased urban population of IMJSHNG over these 40 years. The population density of cities in the semi-arid regions clearly

increased, as illustrated by data from Baotou (station 8), which indicated an increase from 64 persons per square kilometer (km²) in 1949 to 533 persons per km² in 1990 (National Statistical Bureau of China 1990). In 1949, six of the selected stations had populations of less than 10 000, and no station had a population larger than 500 000. But in 1988, five stations had populations of more than 500 000 and none had a population of less than 10 000 people.

Environmental changes

Shrinking lakes. The drier climate, heavy industrial use of water, and increased agricultural and human demands on the water supply, including water dammed for cultivation, have resulted in a reduction in the lake areas in this arid and semi-arid region of China by approximately 6400 km² over the last 40 years. In the 1970s, Luobubo Lake in Xinjiang autonomous region (Northwest China) dried up. Qinghai Lake in Qinghai Province (Northwest China), one of the largest lakes, and the largest salt-water lake in China, decreased in size from 4980 km² in 1908 to 4304 km² in 1986. Data on the larger lakes in Xinjiang, Inner Mongolia and Qinghai are presented in Table 7, which shows the changes in lake area between 1950 and 1980 (Northwest Normal University 1983; Zhao Song-qiau, 1985; Li Jiang-feng, 1990; MCEC, 1991).

Soil erosion. Soil erosion in the Yellow River Valley is increasing. For example, the annual volume of soil loss in Yulin county near the Yellow River was 439 390 000 tonnes in 1979, but in 1985 the silt loss totalled 523 000 000 tonnes.

Sand storms. Strong winds blow in the semi-arid regions in both winter and spring, especially in those areas located close

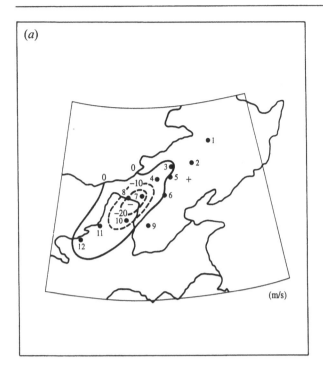

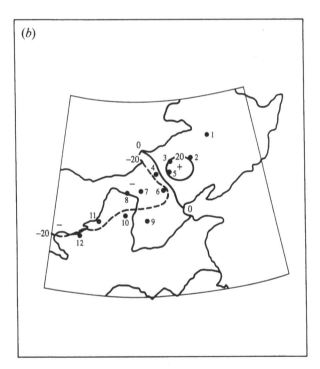

Figure 9: Changes 1950–79 in (*a*) annual mean wind speed and (*b*) number of days annually of strong wind.

Table 6. *Change of grade of population*[a] *between 1949 and 1988 in 12 cities of IMJSHNG*

Year	Number of city[b]											
	1	2	3	4	5	6	7	8	9	10	11	12
1949	4	5	5	5	5	3	3	4	3	5	3	5
1988	3	4	4	4	4	2	2	2	1	4	1	4

Notes:
[a] Grade of population: 1, population > 1 000 000; 2, population between 500 000 and 1 000 000; 3, population between 100 000 and 500 000; 4, population between 10 000 and 100 000; 5, population < 10 000.
[b] See Fig. 2.

drainage control, the problem of salinization of land in the semi-arid regions must be addressed.

Shortage of water resources. The shortage of water resources in the semi-arid regions and the diminishing water table in cities in IMJSHNG is a significant problem.

STRATEGIES FOR THE FUTURE AND SUSTAINABLE DEVELOPMENT

The 'greenhouse effect' could affect the future of the planet. Several of the general circulation model (GCM) studies of climate changes in the semi-arid regions in China are referred to in this chapter (Zhao 1989: GFDL; Manabe and Wetherald 1987: GISS; Hansen *et al.* 1984: NCAR; Washington and Meehl 1984: OSU; Schlesinger and Zhao 1989: UKMO; Wilson and Mitchell 1987). The results of these studies indicate a 3.5–5 °C rise in the temperature in the semi-arid regions, in both winter and summer, resulting from the doubling of carbon dioxide (CO_2) levels. In addition, most parts of IMJSHNG might experience an increase in precipitation of 0.2 mm per day during the winter and a decrease in precipitation of 0.1–0.3 mm per day in the Yellow River Valley. But increased precipitation is expectd for the remainder of the semi-arid regions. Fig. 10 summarizes simulations induced by doubled CO_2 as simulated by the five GCMs. An assessment of GCM simulations of the greenhouse effect in China has also been tested (Zhao Zong-ci and Ding Yihui 1990).

In an attempt to predict the future climate the paleoclimate in the semi-arid regions in China has been investigated. Data obtained from paleo-geological studies and results as simulated by the GCMs have been analyzed and are presented in Figs. 11 and 12 for the year 18 000 B.C.E. (Ice

to the arid and desert regions. Sandstorms and shifting dunes have occurred in the semi-arid areas.

Salinization.   Salinization of cultivated lands in the semi-arid region is becoming a serious problem. Due to the decreasing precipitation, increasing evaporation and lack of

Table 7. *Changes in lake area between 1950 and 1989 in the arid and semi-arid regions of China (km²)*

Regions	Character of lakes	No. of lakes	1950–9	1960–9	1970–9	1980–9
Xinjiang	Arid	16	8832.1	6424.5	4116.5	
Inner Mongolia	Semi-arid/arid	6	3970.7	—	3133.9	2681.5
Qinghai (Qinghai Lake)	Semi-arid	1	4568.0	—		4322.0

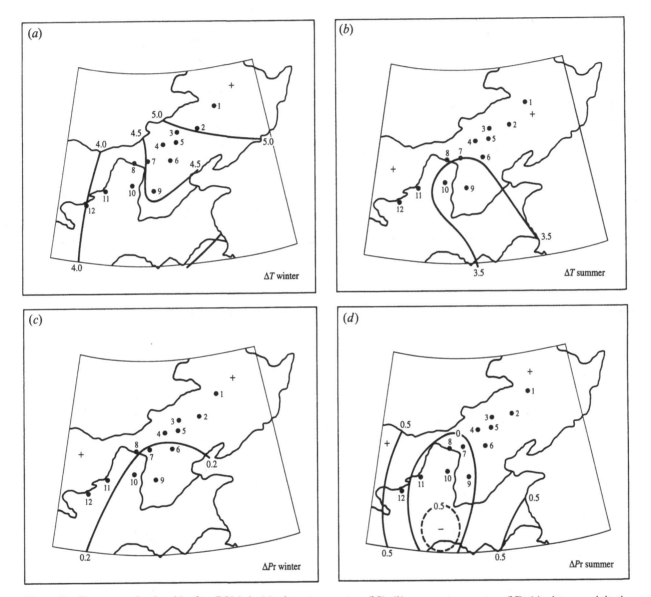

Figure 10: Changes as simulated by five GCMs in (*a*) winter temperature (°C), (*b*) summer temperature (°C), (*c*) winter precipitation (mm/day) and (*d*) summer precipitation (mm/day).

Age) and 6000–9000 B.C.E. (Holocene warm period) (OSU, Gates 1983; GFDL, Manabe and Kahn 1981; UKMO, Mitchell *et al.* 1990: UW, Kutzbach and Gallimore 1989). Figs. 11 and 12 illustrate a 5–10 °C drop in temperature and a drying of approximately 2–2.5 mm per day in the semi-arid regions of China at the time of the Ice Age. During the Holocene warm period the temperature increased by 2–7 °C

and precipitation in the same region increased by 0.1–0.8 mm per day. Over a long period of time, for anomalous cold and warm periods, the warm and wet or cold and dry climates appeared in the semi-arid regions.

From records kept in China over a 500-year period, precipitation records from 16 cold and 16 warm winters have been chosen for study. These statistics are illustrated in Fig.

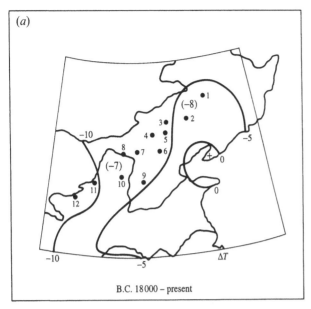

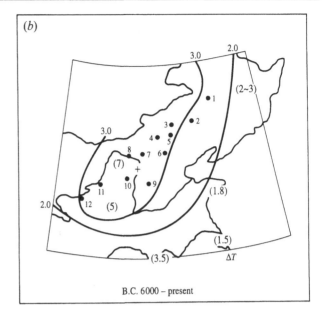

ICE AGE CLIMATE.
Large Numbers: Paleo-geological study (annual value)
Curves: Simulation by GCMs

HOLOCENE CLIMATE.
Large Numbers: Paleo-geological study (annual value)
Curves: Simulation by GCMs

Figure 11: Changes in temperature (*a*) B.C.E. 18 000 to present and (*b*) B.C.E. 6000 to present. Large numbers, paleogeological study (annual value); curves, simulation by GCMs.

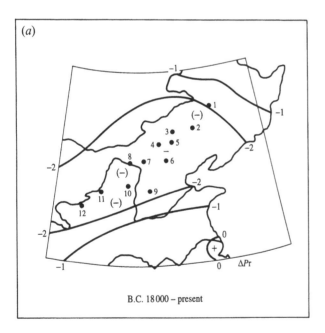

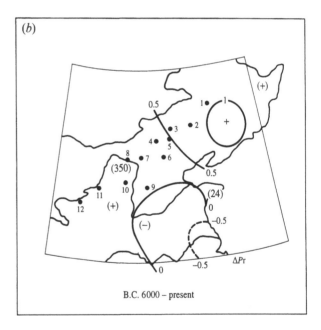

Figure 12: Changes in precipitation (*a*) B.C.E. 18 000 to present and (*b*) B.C.E. 6000 to present. Large numbers, paleogeological study; curves, simulation by GCMs.

Figure 13: Differences in grades of floods and droughts over 16 warm and cold winters during a 500-year period.

13. The records over the last 500 years indicate a weather pattern in IMJSHNG of warm winters and wet summers.

Summaries of data obtained through GCM simulation, paleo-climate analyses and 500-year analyses show that climatic predictions are difficult. It might get warmer and wetter in the eastern and western IMJSHNG in the future, but warmer and wetter or drier in the central part of IMJSHNG (Hetau area). Of course, the patterns of precipitation and temperature in IMJSHNG might be very different over a shorter time scale.

Population and urbanization

Urbanization and immigration have helped to sustain development and progress in the semi-arid regions of China. During the last 30 years the Chinese government has relocated factories, colleges, institutions and people from the industrialized high-density regions to developing low-density regions. Fig. 14(*a*) illustrates this relocation into the semi-arid and arid regions from eastern and central China. Immigrants to this area totalled approximately 25–30 million. Fig. 14(*b*) and (*c*) illustrate the change in density of population and the population growth rate between 1949 and 1985.

Family planning has been a policy in China since the 1970s. According to forecasts made by the Division of Economic Prediction, Center of National Information, by the year 2000 the population of China might total 1.255 billion. Over the last 30 years the population in the semi-arid

regions has accounted for 9.2% of the total population. Therefore, the population in the semi-arid regions might total 115460000 by the year 2000. According to the same statistics, by the year 2050 the population of China might approach 1.5 billion, with a corresponding increase in the population of the semi-arid region of approximately 135000000 to 138120000.

China is urbanizing rapidly. Table 8 compares urban and rural populations for the last 40 years (data from National Statistical Bureau of China 1990; Yang De-qing *et al.* 1990). As shown in Table 1, the population of cities in the semi-arid regions increased very rapidly during that period. Urbanization in China is helpful to family planning and to raising educational levels. Table 9 shows that, in addition, the number of new cities has increased (data from National Statistical Bureau of China 1990).

Education

Sustainable development of the semi-arid regions is dependent upon increasing the educational and cultural level of the population. At present, the educational level of the people in the semi-arid regions is lower than that of East China. Table 10 illustrates percentages of the population attaining specified educational levels from the age of 6 years onwards in this area as compared with the whole of China.

Government at both national and local levels encouraged the more highly educated people in the developed regions to emigrate to the semi-arid regions. At the same time, experts were sent to these regions to teach and to raise the educational level of the local inhabitants. Many young students and workers were sent to the developed regions for study.

Health

Endemic diseases in IMJSHNG include hypertrophy, Kaschin-Beck disease, and fluorine and selenium disorders. Health care networks and health care centers for women and children have long been available in this area to control the outbreak and spread of these diseases, and to improve health conditions in general.

Industry

The semi-arid regions of China are rich in natural resources such as oil, coal and nonferrous metals. The coal reserve in Shanxi Province amounts to about one-third of China's total. In addition, coal deposits in Inner Mongolia and Shaanxi account for one-fourth of the total. Xinjian and Gansu abound in oil deposits. Utilization of these resources is essential for sustainable development.

One of the major coal industries and energy resource bases

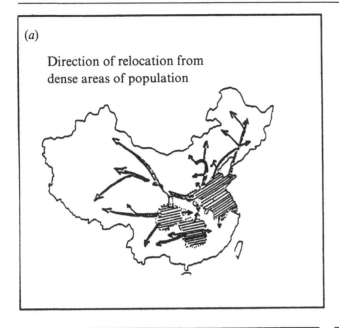

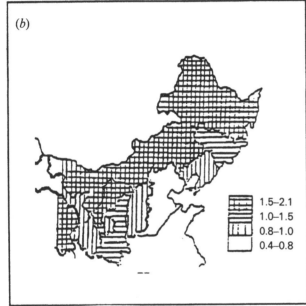

Figure 14: (*a*) Shifts in population, (*b*) rate of population increase 1949–85 and (*c*) changes in population density 1949–85.

in China is in Shanxi province. Three oil and natural gas plants are located in Xingiang, Qinghai and Gansu. The largest hydroelectric power station is also located in Gansu, on the upper reaches of the Yellow River. Two iron and steel plants are located in Shanxi and Inner Mongolia. In addition, the semi-arid region has a metallurgical industry, woollen mills and a sugar-refining industry; these industries will be targeted for development in the future. Fig. 15 shows the locations of some of the industrial bases in the semi-arid regions.

Reforestation

Reforestation has been practiced in the semi-arid regions of China since the 1960s. A forestry plan exists for Northwest China. Shelter belts, windbreak forests, sand-breaks, and shelter belts specifically to protect farmland and river head-waters are already in place, and more are planned for the semi-arid and arid regions. The planting of these shelter belts is a controversial matter, because some scientists argue that the ecosystem balance might be destroyed. Fig. 16 pinpoints the locations of the main shelter belts in the semi-arid regions.

Table 8. *Percentage of urban and rural populations in China*

Year	% of population	
	Urban	Rural
1950	10.6	89.4
1980	19.4	71.6
1987	41.4	58.6

Table 9. *The changes in numbers of cities in the semi-arid regions and the whole of China*

	Semi-arid regions				
	Shanxi	Inner Mongolia	Shaanxi	Gansu	China
1949	2	6	4	1	132
1955	5	6	4	5	164
1965	4	9	4	5	168
1975	7	9	5	4	185
1985	10	16	8	12	324
1988	12	16	11	13	434
1988/1949	× 6	× 2.6	× 2.7	× 13	× 3.3

Table 10. *Percentages of the population attaining specified educational levels from the age of 6 years onwards in the semi-arid regions compared with the whole of China*

Region	> High school		> Middle school		> Primary school	
	%	No.	%	No.	%	No.
Shanxi	9.1	10	33.7	6	77.4	5
Inner Mongolia	9.2	9	31.2	8	68.5	14
Shaanxi	9.8	7	31.5	7	35.2	28
Gansu	7.6	16	21.3	23	52.5	25
Qinghai	6.9	21	23.3	20	53.3	24
Ningxia	7.1	19	25.5	19	55.9	22
Shanghai	25.8	1	56.2	1	83.5	2
Beijing	24.5	2	56.2	1	84.8	1

Irrigation and water conservation

Drought in the semi-arid regions of China is an important issue. Rainfall distribution is uneven, most of the annual rainfall occuring in July and August. Therefore, irrigation and water conservation are very important. Drought-resistant crops have been planted, and irrigation systems have been developed. In addition, groundwater resources have been tapped.

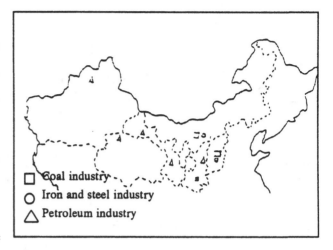

Figure 15: Distribution of the petroleum, coal and steel industries in Northwest China.

Sustainable development of 'green delta' (oasis)

There are oases ('green deltas') in the semi-arid and arid regions in China. By developing these and building shelter belts, water resources have been exploited.

Land improvement, water and soil conservation

Soil management is necessary in the semi-arid regions of China. The salinity and alkalinity of the soil are high for a number of reasons, among which are their high levels in underground water, the dry climate, high evaporation and poor drainage. Improved drainage, irrigation to wash away the salt and alkali, as well as fertilization could reclaim the land.

Most of the semi-arid regions of China are covered by loess. In order to conserve the soil, reforestation, the seeding of grasses, improved irrigation and terracing are being practiced. Engineering surveys have been made.

Desert control and dune stabilization

The semi-arid regions of China are contiguous with desert areas. Deserts, dunes and shifting dunes lie within the semi-arid regions. Shifting dunes have been measured to be moving at between 5 and 20 meters per year in most areas, and 40 meters per year in the flat areas. Towns are in danger of being submerged by sand, farms could be eroded and transportation threatened. In order to sustain development in these semi-arid regions, sand must be controlled. Control of the dunes is important in controlling the desert. Control of dunes, forestation, planting grasses, windbreaks and sandbreaks are all being utilized in desert control. For example, for many years strong winds and blowing sand have damaged farms and taken lives in Minqin, Gansu Province.

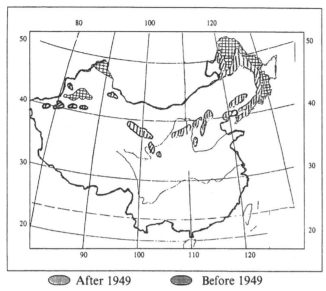

Figure 16: Distribution of forests in North China.
(Note: Some Chinese islands are not shown.)

Over the last 20 years the residents have planted shelter belts, windbreaks and sand breaks totalling 830 km², and built sand dunes and planted grass on 20 million mu (about 1.3 million hectares).

Use of climatic resources

Climatic resources are plentiful in the semi-arid regions of China. These include solar radiation, winds and plentiful sunshine. Sunlight in Northwest Gansu Province totals approximately 3000 hours per year. Solar power and wind power should be harnessed for the development of energy. The hours of available sunlight should also be utilized for the production of crops.

Transportation and communication

Economic and social development are dependent on transportation and communication. A serious lack of both in the arid and semi-arid regions is one of the reasons this area has lagged in progress and development. The area has been isolated for too long, and therefore the development of transportation and communications systems in China is a very important issue. The main form of transportation and communication in the country is the railway, especially for long distances. Other systems include roads, highways, water transport and air transport.

Table 11 shows the development of transport and communications systems in China between 1949 and 1984. As illustrated by Table 11, development took place quickly over this period. It is apparent that development has lagged in comparison with more developed countries, but it must be remembered that China has a large area and a huge population. The development of semi-arid regions of China took place at a more rapid rate than in the eastern part of the country. There were no railways in Xinjiang, Qinghai and Ninxia before 1949. It is noteworthy that since 1949 the length of railroad tracks increased by 14.32% in Northwest China. Fig. 17 illustrates this, as well as the development in inland waterway and air transport in Northwest China from 1949 to 1985. There are many roads and highways in and between the Provinces, although they are not high-speed highways.

Comprehensive utilization

Opinions vary concerning sustainable development in the semi-arid regions, but focus on efficient and improved use of human and natural resources. Table 12 illustrates land utilization in four provinces and autonomous regions in the semi-arid region.

According to an old saying in this region, 'Eat your fill in summer, grow fat in autumn, get thin in winter, and die in spring.' The semi-arid regions contain vast grasslands and grazing lands. Therefore, animal husbandry is significant and plays an important role in this area. But the practice of animal husbandry is weather dependent. Natural grazing lands are unstable, and respond directly to changes in the climate. Excessive grazing and degenerating grasslands pose a serious problem. In order to overcome this problem, natural and artificial grazing lands must be developed for sustainable development. Water resources must be preserved, and a search made for underground water tables. It is important that water conservation projects be established in the semi-arid regions, such as well drilling, the building of dams and reservoirs, and improved irrigation systems. Protecting and reseeding overused grazing lands, and the practice of crop and grazing rotation, are necessary in this area.

There are many hills in the semi-arid regions where fruit could grow. Pomiculture should be developed.

The semi-arid regions of China contain a great deal of arable but uncultivated land. Xinjiang contains 160 million mu, Inner Mongolia 89 million mu and Gansu and Ningxia contain 10 million mu. Farming could be expanded into these areas.

ACKNOWLEDGMENTS

This work was supported by the Chinese Science Committee. I would also like to thank Kui Zhao and Xuejun Zang for collecting data and information.

Table 11. *Development of transportation and communication in China*

Transportation	1949 (km)	1984 (km)
Railway	21 800	51 700
Roads and highways	80 700	926 700
Inland waterways	73 600	109 300
Air		260 200

Table 12. *Land use in some semi-arid regions of China, 1980 (%)*

Regions	Farmland	Forestry	Grassland	Desert	Other
Shaanxi	19.06	22.28	12.64	6.48	39.53
Gansu	7.83	7.22	30.11	14.68	40.16
Ningxia	13.49	2.11	39.69	8.36	36.34
Qinghai	0.81	0.27	56.13	11.88	30.91

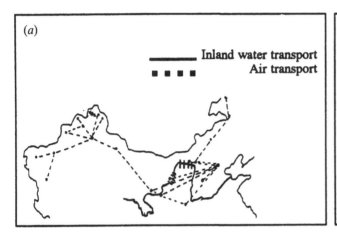
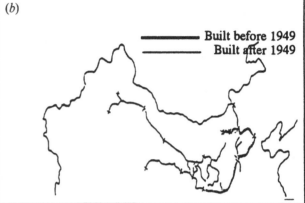

Figure 17: Northwest China's (*a*) inland water transport and airlines and (*b*) railways.

REFERENCES

Academy of Meteorological Science. 1982. *Atlas of Floods and Droughts in China for the last 500 years*. China: Chinese Map Press. (In Chinese.)

China Statistical Information and Consultancy Service Center and International Center for the Advancement of Science & Technology Limited. 1990. *China, Forty Years of Urban Development*. Compiled by the Urban Social and Economic Survey Organization of the State Statistical Bureau. People's Republic of China. (In English and Chinese.)

Chinese Natural Resource Research Committee *et al.* 1988. *Investigations on Natural Resources in the Arid and Semi-arid Regions in China*. China: Science Press. (In Chinese.)

Gates, W. L. 1983. *Ice Age Simulations*. Climate Research Institute Report no. 33. Oregon State University.

Geng Kuan-hong. 1986. *Climate in Desert Areas in China*. China: Science Press. (In Chinese.)

Hansen, J., Lacis, A., Rind, D., Russell, G., Stone, P., Fung I., Ruedy, R. and Lerner, J. 1984. Climate Sensitivity: analysis of feedback mechanisms. In *Climate Processes and Climate Sensitivity*, ed. J. E. Hansen and T. Takahashi, pp. 130–63. Maurice Ewing Series 5. AGU.

Investigative Committee of National Natural Resources of China. 1988. *Investigation on Natural Resources in the Arid and Semi-arid Regions of China*. China: Meteorology Press. (In Chinese.)

Kutzbach, J. E. and Gallimore, R. G. 1989. Pangaean climates: megamonsoons of the megacontinent. *Journal of Geographical Research* **94**:3341–58.

Li Jiang-feng *et al.* (eds.) 1990. *Investigation on Climate, Environment, and Regional Development in the Arid and Semi-arid Regions in China*. China: Meteorology Press. (In Chinese.)

Lin Zhi-guang and Zhang Jia-cheng. 1988. *Climate of China*. China: Shanxi Press. (In Chinese.)

Lu Yu-rong and Gao Guo-dong. 1985. *Atlas on Hydrology and Climate in China*. China: Meteorology Press. (In Chinese.)

Manabe, S. and Kahn, D. G. 1981. Simulation of atmospheric variability. *Monthly Weather Review*, **109**:2260–86.

Manabe, S. and Wetherald, R. T. 1987. Large-scale changes of soil wetness induced by an increase in atmospheric carbon dioxide. *Journal of Atmospheric Science* **44**:1211–35.

Mitchell, J. F. B, Manabe, S., Meleshko, V. and Tokioka, T. 1990. Equilibrium climate change and its implications for the future. In *Climate Change: The IPCC Scientific Assessment*, ed. J. T. Houghton, G. T. Jenkins and J. J. Ephraums. Cambridge: Cambridge University Press.

National Meteorological Center. 1985. *Atlas of Grade of Temperature in China for 1910–1980*. China: Meteorology Press. (In Chinese.)

National Statistical Bureau of China. 1990. *Historical Statistical Data for the Provinces, Autonomous Regions, Cities (1949–1989)*. China: National Statistical Press. (In Chinese.)

Northwest Normal University. 1983. *Atlas of Chinese Nature and Geology*. China: Chinese Map Press. (In Chinese.)

Schlesinger, M. E. and Zhao, Zong-ci. 1989. Seasonal climate changes induced by doubled CO_2 as simulated by the OSU atmospheric GCM mixed layer ocean model. *Journal of Climate* **2**:459–95.

Shao-wu Wang and Zhao Zong-ci. 1981. *Droughts and Floods in China, 1470–1979, Climate and History*, ed. T. Wigley *et al.* Cambridge: Cambridge University Press.

Song Shi-da *et al.* 1980. *Investigation on Droughts in China*. China: Chinese Science Press. (In Chinese.)

Washington, W. M. and Meehl, G. A. 1984. Seasonal cycle experiment on the climate sensitivity due to a doubling of CO_2 with an atmospheric general circulation model coupled to a simple mixed-layer ocean model. *Journal of Geophysical Research* **89**:9475–503.

Wilson, C. A. and Mitchell, J. F. B. 1987. A doubled CO_2 climate sensitivity experiment with a global climate model including a simple ocean. *Journal of Geophysical Research* **92**:13315–43.

Yang De-qin *et al.* 1989. Desertification in China. *Journal of Geography (China)* **5**:7–14. (In Chinese.)

Yang De-qing *et al.* 1990. Paleoclimate in China. *Journal of Geography (China)* **6**:22–30. (In Chinese.)

Zhang Chuen-yuan *et al.* 1983. Changes of lakes in the arid regions. *Journal of Lakes (China)* **2**:9–15. (In Chinese.)

Zhao Song-qiau, Ed. 1985. *Nature and Geology in the Arid Regions in China*. China: Science Press. (In Chinese.)

Zhao Song-qiau, Ed. 1985. *Natural Geography in the Semi-arid Regions in China*. China: Science Press. (In Chinese.)

Zhao Song-san *et al.* 1985. Resources in China. *Journal of Natural Resources (China)* **4**:1–8. (In Chinese.)

Zhao Zong-ci. 1991. Changes of Temperature in China for the Last 40 Years and Urbanization. *Journal of Meteor* 17:4–10. (In Chinese.)

Zhao Zong-ci. 1989. Impacts of climatic change in China induced by greenhouse effect as simulated by the GCMs. (In Chinese.) *Journal of Meteorology (China)* **15**:1–5 (In Chinese.)

Zhao Zong-ci and Zang Xuejun. 1991. Desiccation in China. Abstracts of IUGG XXI, Vienna, Austria, 11 August 1991.

Zhao Zong-ci and Ding Yihui. 1990. Assessment of climate change in China induced by greenhouse effect as simulated by the GCMs. *Journal of Environmental Science. (China)* **2**:74–84.

Zhu Zhen-da, Wu Zhen, Liu Shu *et al.* 1980. *Introduction to Desert in China*. China: Science Press. (In Chinese.)

5: Settlement Advance and Retreat: A Century of Experience on the Eyre Peninsula of South Australia

R. LES HEATHCOTE

INTRODUCTION

This chapter presents ongoing research into the Eyre Peninsula of South Australia, where there appears to have been at least a century of relatively well-documented intensification of land use paralleled by fluctuating socioeconomic fortunes and locally devastating impact upon the ecosystems of the area.

Among the current scenarios of future climate change for Australia are forecasts of the reduction of winter rainfall of up to 20% and increases in summer rainfall of similar proportions. Associated with these changes are increases in annual average temperatures of 1–3 °C (Pittock and Hennessy 1991). Such forecasts have caused concern among rural planners in the cereal grain producing areas of southern Australia, especially where current economic constraints are reinforcing the marginality of much rural production and rural life-styles.

What follows is a preliminary account of research focusing upon the characteristics of the Eyre Peninsula (Fig. 1), the sources available for its study, and suggestions of some preliminary findings and general questions.

THE EYRE PENINSULA STUDY SITE

Sources

In terms of the last 100 years of human occupation, the Eyre Peninsula offers both abundant and spatially detailed data sources. The process of European occupation of the British Colony of South Australia, from 1836 onwards, was undertaken by a civilian government concerned with providing an orderly framework for the occupation and exploitation of the natural resources by the application of capital and labor. The intention here, as elsewhere in Australia, was the encouragement of private development within a broad framework

of official land survey but private land ownership (Williams 1974).

The result was a survey system where individual farm holdings were defined and statistics collected within a framework of standard spatial units. 'Hundreds' (approximately 100 square miles or 259 square kilometers) and 'Counties' (combinations of Hundreds) were surveyed in advance of agricultural occupation. Statistics on actual areas occupied and the production are therefore available for most of the last 100 years on a constant spatial grid which lends itself well to GIS (geographical information system) analysis. In addition, the existence of the parliamentary democracy and a propensity to appoint special committees of enquiry or Royal Commissions to investigate societal problems as and when they occurred, has provided a rich archive of the debates and philosophical contexts for resource management over the last century.

Physical characteristics

The Eyre Peninsula (almost 49 000 square kilometers) is a climatic transition zone from a 'Mediterranean' type climate with winter rainfall averaging approximately 500 millimeters (Humid Mesothermal Dry Summer, 'Csa') to semi-arid steppe climate with highly variable rainfall averaging approximately 250 millimeters (Dry Steppe, 'Bsh') (Trewartha 1964). Extensive plains with quaternary linear sand-dune systems are broken by granite inselbergs and relict salt lakes over sparse and often brackish aquifers, contained in limestones partly covering the Pre-Cambrian and locally lateralized and deep weathered basement rock complex (Parker *et al.* 1985). At the time of first European contact the area was covered by a range of vegetation, from woodlands including Red Gum (*Eucalyptus camaldulensis*) and Peppermint Box (*E. odorata*) in the south, through dense scrubs of Mallee (*E. socialis, E. diversifolia* and many variants) with

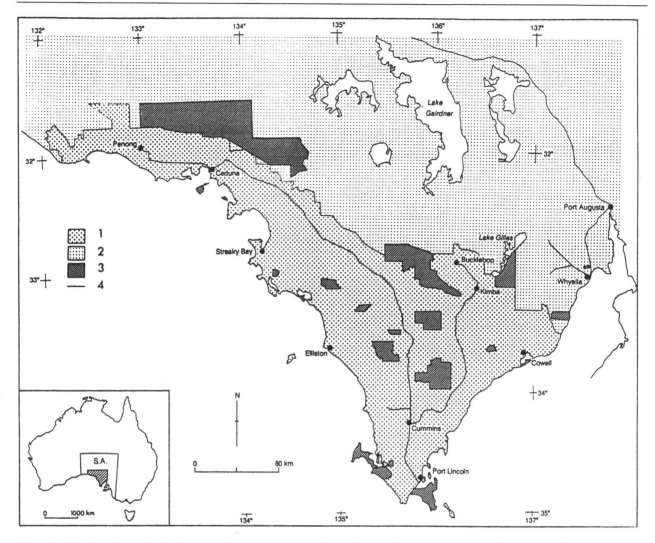

Figure 1: The Eyre Peninsula, South Australia. 1, agricultural land (cropland and improved grazing land); 2, pastoral land (unimproved grazing land); 3, reserved land (national parks and scientific reserves); 4, railways. (Source: Map of 'South Australia General Reference 1:500 000', Department of Lands, Adelaide, October 1990.)

Pines (*Callitris* spp.) and Tea Tree (*Melaleuca* spp.) dominating the center, to low woodlands of Mulga (*Acacia aneura*) and Western Myall (*A. sowdenii*) merging into Saltbush (*Atriplex* spp.) and Blue Bush (*Maireana* spp.) shrubland on the northern fringes of the desert interior (Lange and Lang 1985).

Land settlement history

Prior to European intrusion in the early nineteenth century, the area was supporting approximately 2000 semi-nomadic Aborigines relying upon hunting, fishing and food collection. While there was some evidence of inter-group conflicts, the main conflict came with the European invasion, which reduced the cultures to a minority role as the result of a combination of physical conflicts, disease and social disrup-

tion (Berndt 1985). By 1986 the population claiming Aboriginal descent was 1232 from a total population of 33 644.

Originally considered as a possible main base for European colonization of the new British Colony of South Australia in 1836, Port Lincoln was rejected because of an uncertain water supply and indifferent soils, in favor of the present State capital, Adelaide. Those two factors have continued to plague European settlement on the Eyre Peninsula ever since.

Initial European settlement favored the open woodlands on the limestone plains of the western peninsula and hills of the northeast, where natural grasses and some access to surface water or shallow wells provided support for sheep and cattle raising upon land held under terminating pastoral leases. By 1888 when most of the leases expired, the colonial government was pressured to open the land for agriculture,

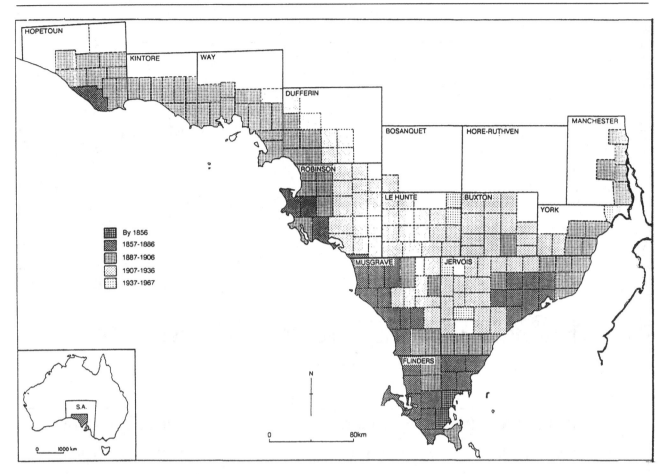

Figure 2: The spread of agricultural settlement on the Eyre Peninsula. The key indicates the dates of official opening of the Hundred survey units for agricultural occupation. While the land was not always immediately occupied, the sequence does indicate the broad historical trends of agricultural advance.

as techniques for clearing the relatively untouched mallee areas had been developed elsewhere in the colony and settlers retreating from the northern, more arid frontier were clamoring for new lands to develop (Meinig 1962). The resultant story is outlined in Figs. 2 and 3 and the following discussion is intended as commentary upon those figures.

The agricultural occupation of the pastoral leases and adjacent mallee areas began in the 1890s, but had to rely upon uncertain water supplies for domestic and livestock use and had to get produce to the coast for transport to the Adelaide or expanding overseas markets. Effective occupation of the mallee interiors had to await two infrastructure developments. First was the construction of a narrow-gauge railway along the axis of the peninsula, begun in 1905 (Port Lincoln to Cummins) and completed to Ceduna by 1915, Penong by 1924 and Buckelboo by 1926. The second was the construction of a water pipeline reticulation system from a catchment in the south along the railway axis to the north and west, begun in 1916 and basically completed by 1929 to Ceduna. This complemented the earlier rainfall runoff catch-

ments around most of the granite inselbergs and the corrugated iron roof catchments built by the government upon the farm blocks prior to occupation.

The advancing wave of agricultural settlement paralleled the railway and water pipeline construction from south to north (see Fig. 2). The mallee was rolled down and then burned, with the first grain crops sown into the ashes, using the locally developed stump-jump ploughs and reapers to help sow and bring home the harvest. Land clearance for agriculture increased rapidly through the period of the 1910s to the 1930s and revived in the western areas in the 1950s to the 1960s. By the 1970s, 63% of the original vegetation had been removed and, although cereal and livestock production was impressive, there was evidence of accelerated soil erosion and long periods of economic stresses had led to farm amalgamations.

Currently the low wheat and wool prices together with droughts are reviving conditions similar to previous periods of stress and there are fears for the future viability of rural settlement in the area. Fears of further reduction of the

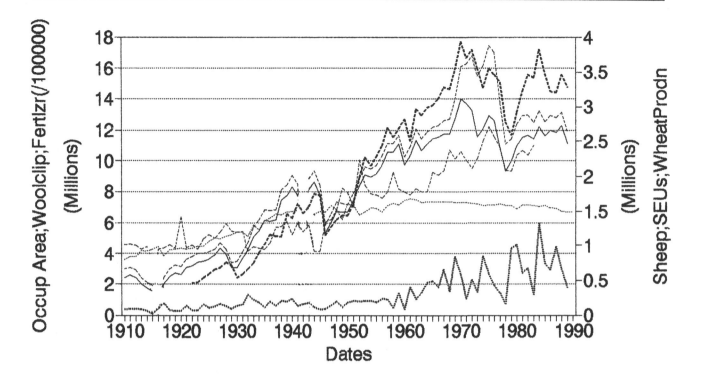

Figure 3: Eyre Peninsula production 1910–90. Occupied area (ha), annual totals of land in private ownership; Sheep, annual totals of sheep grazed on the peninsula; SEUs, annual total of sheep-equivalent units (sheep + cattle × 8); Wheat prodn (t), annual wheat production in tonnes; Woolclip (kg), annual woolclip in kilograms; Av. fertilizer (kg/ha), annual total of superphosphates applied. (Sources: South Australia Statistical Registers and Australian Bureau of Statistics Data.)

winter rainfall as part of the 'Greenhouse effect' have intensified local, regional and eventually national concerns and led to the first attempt in Australia to bring together local settlers and administrators, academics, conservationists and state and federal governments in a cooperative approach to regional planning: the Eyre Peninsula Project. To date implementation of the project has been slow but some progress has been made (Smailes and Heathcote 1992). One of the problems facing the project is the measurement of the extent of desertification occurring on the peninsula.

DESERTIFICATION: THE PROBLEM OF DEFINITIONS

That desertification was occurring on the Eyre Peninsula was suggested by the United Nations World Map of Desertification on which the peninsula was identified as an area with a high risk of desertification and the surface subject to salinization or alkalinization (UN 1977). Although subsequently criticized in terms of its methodology (Carder 1981), the map basically reinforced the evidence accumulated as part of the national soil conservation survey of 1978 (RDDEHCD 1978). These were rather general measures. What, however, was meant by desertification and what was the more detailed evidence?

Desertification

A recent review of the evolution of meanings associated with the term 'desertification' has been provided by Odingo and his final suggested definition is close to that accepted by the Ad hoc Committee on Assessment of Global Desertification on 16 February 1990 in Nairobi (UN 1990:28):

Desertification, or land degradation, is the process of land degradation characteristic of the arid, semi-arid and sub-humid areas of the world. The process, which is largely human-induced, is normally exacerbated by adverse climatic conditions such as prolonged drought or desiccation, which enables it to strike at the land resource base by weakening the physical, biological and economic potential of the land, thereby severely reducing or curtailing overall productivity. (Odingo 1990:47)

Documentation of this process, so defined, therefore requires evidence of the adverse climatic conditions (prolonged drought or desiccations) and the reduced or curtailed overall productivity to be measured in some way.

Drought

The problems of defining drought have been identified previously (Heathcote 1969, 1988) and for the purposes of this chapter, only the evidence for meteorological droughts (using the Commonwealth Bureau of Meteorology definitions) and agricultural droughts (using agronomic definitions oriented to cereal cropping) will be considered.

Desiccation

In his discussion of desertification Odingo suggested that Hare's (1987:6) concept of 'true desiccation – a prolonged period in which drought slowly and intermittently intensifies' over a period of decades, was a useful component in the analysis of possible linkages between drought and desertification (Odingo 1990). The cumulative deviation of rainfalls from the long-term mean may be one way of identifying such periods.

DOCUMENTING DROUGHT ON EYRE PENINSULA

Two measures of drought appear to be relevant here, namely meteorological drought (as defined by the Bureau of Meteorology using rainfall deciles) and agricultural drought defined in terms of rainfall inadequate for crops to mature.

Meteorological drought: patterns in time

Using the annual maps of rainfalls in the first and second deciles available in Gibbs and Maher (1967) and subsequently in the Commonwealth Bureau of Meteorology Monthly Rainfall Reviews, the percentages of the Eyre Peninsula (defined as the area within the counties) so affected were tabulated and the years with droughts were ranked according to area affected (Table 1). Finally, the percentages

Table 1. *Meteorological drought on Eyre Peninsula*

Area in 1st decile (%)	Years	Areas in 1st + 2nd decile (%)	Years
100	1928, 1982	100	1888, 1897, 1928, 1929, 1957, 1982
90	1897	87	1967
78	1967	79	1965
70	1914	78	1914
64	1965	76	1918
		73	1961
No other		66	1977
years over		61	1988
50% in 1st		55	1891
decile		54	1894
		52	1943

Sources: Gibbs and Maher (1967); Comm. Bur. Met. Monthly Rainfall Reviews 1967–90.

Table 2. *Drought decade rankings for Eyre Peninsula 1890s–1980s*

Rank	Cumulative area in 1st decile		Cumulative area in 1st + 2nd decile	
1	1960s	(169)	1920s	(254)
2	1920s	(124)	1960s	(239)
3	1890s	(110)	1890s	(219)
4	1980s	(100)	1910s + 1980s	(161 each)
5	1940s	(73)	1940s	(159)
6	1910s	(70)	1950s	(133)
7	1950s	(59)	1880s	(122)
8	1880s	(28)	1970s	(94)
9	1970s	(8)	1930s	(61)
10	1900s	(3)	1990s	(42)
11	1930s	(0)		

Sources: Gibbs and Maher (1967); Comm. Bur. Met. Monthly Rainfall Reviews 1967–90.

in both first and second deciles were summed for each decade and the decades ranked according to apparent extent of drought occurrence (Table 2).

Table 1 shows that the whole peninsula was drought affected (rainfalls in the first decile) on only two years (1928 and 1982) between 1885 and 1990, but if rainfalls in the second decile are included, the total moves to six years (1888, 1897, 1928, 1929, 1957 and 1982). The table also shows that over half the peninsula experienced rainfalls in the first decile in six years over the period, while if the areas affected by rainfalls in the second decile are added, the peninsula had 16 years so affected.

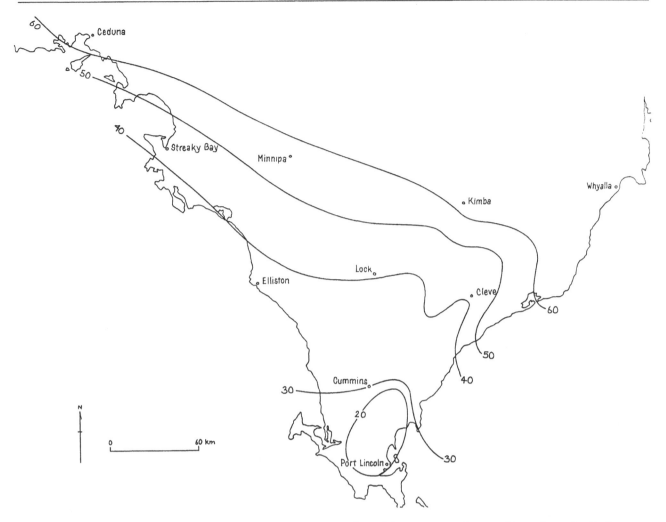

Figure 4: Agricultural drought frequency on Eyre Peninsula. 20–60, isolines of the percentage of years with continuously effective rainfall (for cereal crops) for fewer than 5 consecutive months. (Source: Trumble 1948.)

Table 2 shows that for first decile rainfalls, the worst four decades were the 1960s, 1920s, 1890s and 1980s respectively, while for combined first and second decile rainfalls the worst four decades were the 1920s, 1960s, 1890s, and 1910s and 1980s respectively. Finally, summing the combined percentages for 11-year sequences, the periods 1957–67 with a score of 372 and 1888–98 with a score of 319 stand out as times of extensive and persistent meteorological droughts.

While these are admittedly crude measures, they do provide the general background against which to examine other evidence of droughts and their impacts, and if a causal link between drought and desertification is postulated, we might look particularly closely at these periods for evidence of desertification processes at work.

Agricultural drought: a spatial gradient

The meteorological drought data above assumed an even potential for drought occurrence across the peninsula, but if agricultural drought is considered then a gradient of vulnerability from north-northeast to south-southwest across the peninsula can be shown. In 1948 Trumble published maps of drought frequency for South Australia showing 'the number of years in a hundred in which the season of continuously effective rainfall [for cereal grains] is less than five months [the critical limit for grain to mature]' (Trumble 1948:58).

For the Eyre Peninsula the pattern showed a gradient from over 60% frequency/probability in the north-northeast to less than 20% frequency/probability in the south (Fig. 4). This same basic trend is evident in the map of the index of drought incidence (using percentiles) in the *Year Book of Australia* (1984:34). Thus it would appear that the increased frequency and liability to agricultural drought as one moves north through the peninsula would be a factor in the potential for desertification. Indeed, if productivity were to be a criterion, Trumble had already identified 'marginal wheat-growing areas, including such centers as . . . Cowell, Ceduna . . . Kimba . . . there are, on the general average, in these areas, about two drought years in three' (Trumble 1948:58–9).

In such areas crop failure might be one indicator of potential desertification.

DOCUMENTING DESICCATION ON EYRE PENINSULA

While the above data identify years and decades as droughty, there is little sense of any trend over time – the vital ingredient of Hare's concept of desiccation. Prior studies, however, do provide evidence of rainfall trends over the last 100 years and the relevant information for the Eyre Peninsula is summarized below.

Using the residual-mass concept for monthly rainfalls, Foley (1957) showed that the South Australian agricultural areas of the 1950s experienced an upward trend in rainfall from 1900 to the mid-1920s before a long decline to the mid-1940s and generally stable figures in the 1950s. The peninsula appeared to be relatively unaffected by the shift in climatic belts 1881–1910 to 1911–40 as depicted by Gentilli (1971:206), but both Pittock (1981) and Russell (1981) suggested that there were observable trends in rainfall over the peninsula. Pittock suggested that there was an increase of 0–50 millimeters in mean annual rainfall between 1913–45 and 1946–74, and Russell suggested that the period 1925–50 had winter rainfalls generally below the long-term means and that there seemed to be significant discontinuities in the residual-mass curves in 1903–5, 1927 and 1950–51.

More recently Hobbs suggested for the period from 1891 to 1982 in South Australia: 'The overall period to the mid-1920s can be characterized as relatively wet, the period from the mid-1920s to the late 1940s as relatively dry, and the period since then as one of greater variability with shorter wet and dry spells' (Hobbs 1988:295). In terms of the cereal grain growing season rainfall, Hill (1987) showed a division between the northern and southern halves of the peninsula, with rainfall decreases from the 1930s–1940s in the south and increases in the north. This trend was generally confirmed by Burrows (1989) for the period 1937–87, with decreases of up to 10% from long-term means observed.

In summary, it would appear that:

(1) There have been observably different trends in rainfall over the Eyre Peninsula since the last decade of the nineteenth century.
(2) Those trends seem to have taken the form of gradual increases over the period from the 1880s to 1920s; followed by decreases from 1920s to the 1940s; with separation of the decreases in the south from increases in the north thereafter. The potential, therefore, for the processes of desiccation and desertification to be identified in the area after

the 1920s would seem to be high. How does the evidence of desertification measure up?

DOCUMENTING DESERTIFICATION ON EYRE PENINSULA

Given the general definition of desertification adopted here, any attempt to provide documentation must consider a variety of potential indicators. Briefly, these might include evidence of:

(1) changes in available biomass over time;
(2) accelerated soil erosion;
(3) increased soil salinity;
(4) declining agricultural productivity (however measured);
(5) declining population carrying capacities.

A sample of the evidence for each will be provided to illustrate the potentials and some of the questions they pose for an understanding of cause and effect.

Changing biomass

As mentioned earlier, European settlement from the 1880s onwards has been accompanied by the widespread removal of the original sclerophyll woodland, mallee scrub and saltbush and their replacement by annual grain crops or long rotations of pastures and grains. By 1975, 62.8% of that original vegetation had been removed within the counties and, although it was claimed at the time that the most suitable agricultural lands had already been cleared, clearance of vegetation continued, until by 1988, 71.5% had been removed (SAICVC 1976; EPCSA 1988).

As noted earlier, the periods of most rapid clearance appear to have been the 1910s to the 1930s and the 1950s to the 1960s, periods of the rapid increase of farmland under official financial incentives (Boeree 1963; Williams 1974). Both these periods were times of declining annual rainfalls, and extensive meteorological droughts. Comparing the biomass of the original vegetation with that of the cultivated paddocks under drought conditions there would be no dispute about reduction in biomass, but good seasons might encourage a debate as to which was the larger per unit area.

Accelerated soil erosion

Evidence of accelerated soil erosion on the peninsula over the last 100 years is of varying quality but there is no doubt that accelerated erosion did take place at identifiable times and locations. The evidence, however, must be set in the context that approximately 56% of the surface comprises dunes and swales formed some 3000 to 6000 years ago in a previously more arid climate, and that these areas were naturally

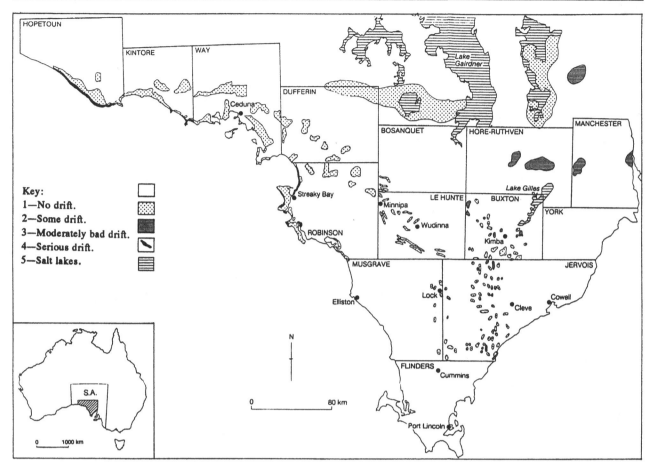

Figure 5: Soil erosion on Eyre Peninsula 1938. (Source: SAPP 1938.)

vulnerable to erosion once the protective original vegetation was removed. Both Crocker (1946:91) and Laut *et al.* (1977:3–4) commented on the vulnerability of the dunes to reactivation once cultivated.

Official concern for accelerated soil erosion in South Australia as a whole had been evident as early as 1867, and by 1899 the Minister of Agriculture was suggesting tree planting for soil control and contemplating 'legislation to penalize owners of drifting land' (Williams 1974:303).

For the Eyre Peninsula, however, the first official inquiry reported in 1938 as part of a state-wide survey (SAPP 1938). The map with the report showed that the peninsula was not the worst affected area, although it did have a band of small eroding areas from northwest to southeast across the center of the peninsula (Fig. 5), which broadly coincided with the dune areas identified by Crocker (1946) and the calcenite dunes and plains of Laut *et al.* (1977). While most of the eroding areas had only 'some drift' – 'small patches still bare of growth and drifting with every wind,' the naturally vulnerable nature of the sandy soils was noted by the Committee and, recognizing that 'in many cases' erosion was 'a measure of mismanagement in the past as much as the

natural liability,' they noted that 'Some of this uncleared land is so very sandy that it is to be hoped that the Government will never be tempted to open it for [agricultural] selection but will retain the areas as fauna and flora reserves' (SAPP 1938:16). As we have seen, however, this was a vain hope.

Reports of erosion continued through the 1940s (SAPP 1946:No. 62) but were surprisingly relatively muted through the 1950s and 1960s. The 1959 'major drought' brought 'very few wind erosion problems in upper Eyre Peninsula,' but the 1967 dry season brought reports of soil drift in the central north of the peninsula. Much of this was 'associated with extensive areas of newly cleared land' (Wetherby *et al.* 1983:11).

The next major official concern came in 1978 when a combination of drought and reduced production occurred after a series of years of low prices and farmers attempted the traditional strategy of 'cropping themselves out of a drought' – i.e. sowing a larger area to provide a larger financial return. The results are set out in Fig. 6 and Table 3.

Once again, as in 1938, the definitions of the types of erosion were less than precise, but the area of accelerated

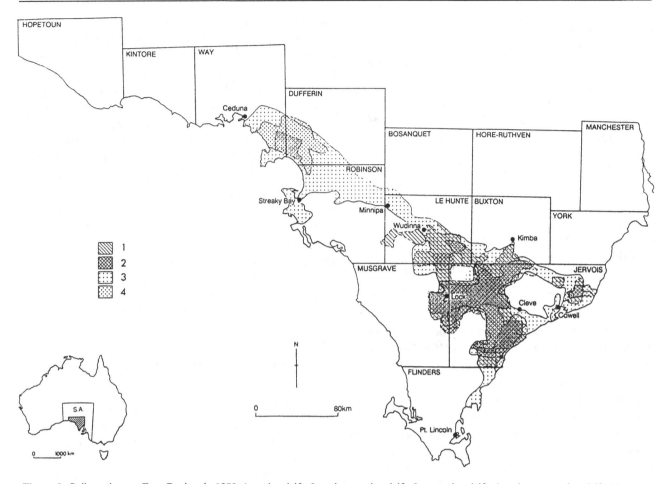

Figure 6: Soil erosion on Eyre Peninsula 1978. 1, active drift; 2, serious active drift; 3, sweeping drift; 4, serious sweeping drift. Note that rehabilitation of areas keyed 1 and 2 is estimated to take 4 years; rehabilitation of areas keyed 3 and 4 is estimated to take 1 'normal' year. (Source: Wetherby *et al.* 1983.)

erosion so defined extended from northwest to southeast across the center of the peninsula, again coinciding well with the dune systems identified in Crocker and Laut *et al.* Dry seasons in 1975–6 and 1976–7 together with drought in 1977–8 were blamed, together with 'an ill-defined start to the 1977 season, water repellence on non-calcareous sandy soils, overstocking, excessive cultivation associated with the use of pre-emergent herbicides and unsuitable cereal cultivars for the sandier soils' (Wetherby *et al.* 1983:1). More specifically, 'The situation in 1977 was aggravated by a large increase in area sown to crop as farmers sought to recoup losses from the previous dry seasons. In addition farmers were reluctant to quit stock until absolutely necessary because very high prices were likely to follow the end of the drought' (Wetherby *et al.* 1983:2). In-drought prices of $1 per head were compared with post-drought prices of $25 per head.

By 1980, however, it was claimed that there was 'little evidence of the effects of the prolonged drought' as a result of more favorable seasons and extensive government drought relief (Wetherby *et al.* 1983:40). The economic costs of

rehabilitation were nonetheless significant. Some 937 (42.3%) of the 2187 total rural landholders on the peninsula had applied for drought relief, of which 858 were given loans totalling $13.4 million – some 64% of the state's Drought Relief payments for the period September 1976 to June 1978 (Ellis and Porter n.d.:50–1).

In a more general study of wind erosion in Australia 1960–84, Burgess *et al.* (1989) provided a map of excess wind erosion (i.e. that considered to be above the climate wind erosion potential). On the Eyre Peninsula, the Ceduna district showed excessive dust storm frequency in the moderate to high class compared with the remainder of the peninsula. They explained this localized occurrence as 'associated with the destabilization of calcareous coastal sand dunes by rabbits, overgrazing and recreational activities,' and explained the contrast between the Ceduna area and Streaky Bay to the south as 'a function of the wind erodibility of the soils' (Burgess *et al.* 1989:106). Probably a more significant factor was the rapid expansion of the cultivated area in the County Hopetoun, west of Ceduna, in the period. For

Table 3. *Desertification in 1978 on Eyre Peninsula*

Production trends	1974/5	1975/6	1976/7	1978/9	1978/80
Cereal production					
Crop area (000 ha)	697	582	633	781	880
Yield (000 t)	779	621	434	219	1 299
Yield (t/ha)	1.1	1.1	0.7	0.3	1.5
Sheep-equivalent					
units (000)	2 907	2 858	2 243	1 777	1 886

Drifting areas	Dec. 1975	May 1978	Aug. 1978	Mar. 1979
Active (ha)	500	27 500	6 000	2 000
Sweeping (ha)	10 100	138 000	50 000	5 000
Total (ha)	10 600	165 000	56 000	7 000

Source: Wetherby *et al.* (1983).

example, in 1955 some 1836 hectares had been cultivated, by 1965 this had risen to 3936 hectares, by 1975 to 4006 hectares and by 1985 to 10 294 hectares. New land clearance may have destabilized the local sandy soils.

Local government records offer evidence on a variety of matters relevant to drought, desiccation and desertification. Some of the district clerks or chief executive officers included a summary of seasonal conditions in their end-of-year statements and these provide a 'grass roots' view of the weather patterns over the past year and their significance to the local community. These may help identify critical thresholds of stress for the communities. Accelerated erosion shows up in the concern for the condition of the local roads, particularly when blocked by soil drift or water erosion. Such references are frequent, not only because of the local government responsibility for road maintenance, but also because of the use of road maintenance as public relief work as part of traditional drought relief measures. The references may be plotted over time and give some indication of relative frequency of concerns (Fig. 7).

Briefly, relief funds could be used to employ farmers on road maintenance, and the councils often allowed farmers who were behind with their rates to work off their debt similarly. The tasks were usually described as either grubbing (i.e. clearing vegetation off the surveyed road alignment), scooping (i.e. the removal of drifted soil which was blocking the roads) or rubbling (i.e. the laying down of road metal over irregular or soft sandy road surfaces to provide improved all-weather access). Interestingly the incidence of grubbing, high in the 1920s on the peninsula, declined over time and was replaced by scooping and rubbling through the 1930s and 1940s as the incidence of drift increased.

In addition, the minutes of local government administration give clues to the actions which may have exacerbated

the erosion problem. To cope with drift on to roads, for example, some councils encouraged and even required the removal of existing natural vegetation to keep the soil moving on: 'That [name] be written suggesting that he remove all timber on both sides of the road near his section to allow sand drift which is coming from his land to blow away naturally as this sand is making the road nearly impassable' (Kimba District Council Minutes, 11 February 1927).

Such strategies were relatively common into the 1940s. The same council in 1984 had a more positive attitude to some of its continuing drift problems: 'That Buckelboo Football Club be given permission to remove the drift soil from the road reserve adjacent to sections 13, 14 and 16 Hd [Hundred] of Pinkawillinie to top dress the Buckleboo Oval' (Kimba District Council Minutes, 12 June 1984). Considering that the soil probably contained fertilizers from the adjoining paddocks, it was probably a good investment.

Salinization of land

Laut *et al.* commented upon the naturally saline nature of parts of the peninsula:

> Salt accumulation commonly occurs in low lying areas, especially where there is no through-going drainage, and there are many salt lakes in the province. The main source of salt is sea spray blown inland, deposited on soil or vegetation, and carried to these low-lying areas by surface run-off or sub-surface drainage. (Laut *et al.* 1977:4)

In 1982 the map 'Salting of non-irrigated land in Australia' showed portions of the southern and eastern peninsula affected by induced saline seepages (the result of the effects of cultivation) and 20 000 hectares were estimated to be affected, with implications of reduced water quality for domestic use (WPDSA 1982:64). Some scalding was also noted for the Ceduna–Streaky Bay areas. In addition Australian Bureau of Statistics data showed that 53·148 hectares of land was saline and had been lost to cultivation by 1984–5 (ABS 1986–7:Cat. 7113.4).

Sampling of the local government sources so far has not found much direct evidence of salinity problems, but one exception would be a request for reduction of rates on the grounds that 'This land is not freehold, there are about 500 acres of granite outcrop on it which is valueless and about 100 acres of salt rises which are increasing' (Kimba District Council Minutes, 9 October 1931).

Decreasing productivity

Lack of space and the incomplete nature of the research prevent anything other than a brief review of some measures here. Fig. 3 illustrates some livestock and wheat production

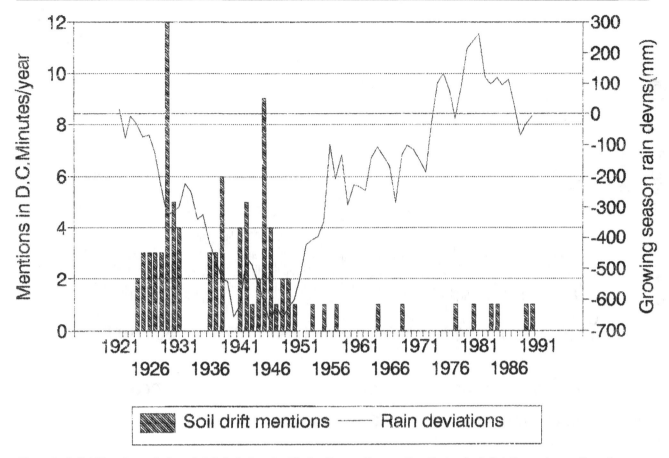

Figure 7: Soil drift and cumulative rainfall deviations for Kimba, Buxton County, Eyre Peninsula. Soil drift mentions, soil erosion problems noted in the monthly District Council meetings and totalled annually; Rain deviations, cumulative deviations of annual growing season rainfalls (April–October) from the long-term mean. (Source: Minutes of the Kimba District Council 1925–91 and Commonwealth Bureau of Meteorology records for Kimba.)

trends which show increases through time, but at differing rates and with periodic reversals. From 1910 onwards the expansion of the sheep flock was halted in 1915, 1929, 1946, 1960, 1971 and 1978, mainly by droughts but more recently by a switch to cattle in the 1970s as a result of improved beef prices. The woolclip has generally followed the sheep trends as might be expected. Wheat production has increased but become more variable since the 1960s despite increased application of fertilizers.

Decreased rural population carrying capacity

The rural population on Eyre Peninsula has reflected the size and number of agricultural holdings over the years. These in turn have reflected first the process of intensification of land use over the last 100 years – being the change from extensive pastoral properties to smaller grain–livestock farms; and second the process of farm amalgamations and restructuring to achieve economies of scale, when the best lands had been developed and expansion into poorer lands began to be shown to be risky. This restructuring was partly market-driven, but also partly officially blessed, as through the Marginal Lands Scheme of the 1940s to 1960s and the various Rural Assistance Schemes in operation since.

On the peninsula the number of rural holdings peaked in 1938 at 4015 and by 1990 had fallen to 1731; the area in farms had, interestingly, also diminished from a peak in 1961 of 7.5 million hectares to the current 6.9 million, reflecting the abandonment of some farming lands and the conversion of others to public reserves (Fig. 8). The total population on the peninsula, as we have seen, however, had risen from approximately 2000 in 1836 to 33 644 at the 1986 census. This increase in overall population has been urban in nature, and specifically in the larger towns, as has been the picture for rural Australia generally.

This picture of rural population increase followed by decline, alongside overall population increase, is paralleled in the local government sources by evidence of the creation of new local government areas, the growth in number of rate payers and the clearing of new roads, followed by evidence of the closure of unused road reserves, reduction in size of surveyed townships, reduction in number of rate payers,

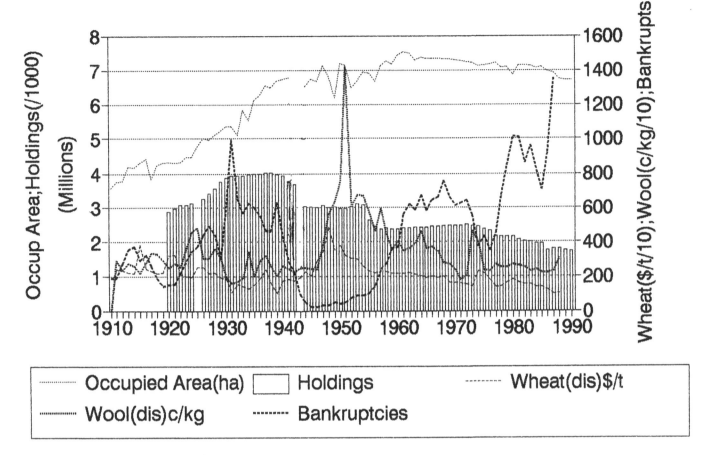

Figure 8: Eyre Peninsula production contexts 1910–90. Occupied area (ha), annual totals of area in private ownership; Holdings, annual totals of agricultural holdings; 3, Wheat (dis) $/t, discounted wheat price in Australian dollars per tonne; Wool (dis) c/kg, discounted wool price in Australian cents per kilogram; Bankruptcies, registered bankruptcies and insolvencies in South Australia (both rural and urban enterprises). (Sources: South Australian Statistical Registers and Year Books.)

followed by attempts to increase rate payers by enlarging the local government areas, to sell lands and town lots to repay arrears of rates, down to the symbolic removal of the last street light at Buckleboo (Kimba District Council Minute, 13 November 1990).

The economic dimension

Changing productivity and rural population carrying capacities, however, need to be set in the changing economic context, and this can be best encapsulated in two sets of statistics. First, in Fig. 8, are provided the trends of actual and discounted prices for wheat and wool – the two main rural land products. The disastrous fall of the discounted value of the produce through the 1980s has exacerbated the plight of farmers trying to cope with heavy debt burdens and with interest rates on their debts which have doubled to over 20% during the decade.

A second series of statistics are the estimates of the volume of production of wheat needed per hectare to cover the costs

of its production (i.e. before a profit on the production could be achieved). In the 1880s this break-even point was thought to be approximately 0.2 tonnes per hectare; by the 1930s it was 0.4 tonnes per hectare; by the 1960s it was 0.6 tonnes per hectare; by the 1980s it was 0.8 tonnes per hectare; and currently it is close to 1 tonne per hectare. However, in the decade of the 1980s, of the 148 Hundreds on the peninsula producing wheat for over half the period, one-third had yields of less than 0.8 tonnes per hectare.

Given the pressures reflected in these two sets of statistics, it is questionable whether much of the farming activity is sustainable without some improvement in the economic climate.

SUMMARY AND QUESTIONS FOR FURTHER CONSIDERATION

This survey of the documentation of drought and desertification on the Eyre Peninsula has only hinted at the array of potential sources and the scope they offer for analysis not only of some of the physical causes, but also of the human activities which affect the desertification processes and the question of sustainable resource management. The physical processes of desertification have been shown to reflect drought, the incidence of desiccation, and the existing soil characteristics, but also farm management practices, contemporary economic conditions, government policies and contemporary perceptions of what was happening to the landscape in times of physical, economic and social stress.

Despite the early stage of the research, several questions have been raised in collaboration with a colleague (Peter Smailes) from the University of Adelaide with whom the author is collaborating on the Eyre Peninsula Project, and we offer these observations in conclusion:

(1) We agree with the findings of Lowrence et al. (1986) that the question of sustainability needs to be addressed at various levels, each of which is linked to the adjacent levels of activity. Thus the 'macro-economic sustainability of an entirely agriculturally based region', such as the Eyre Peninsula, can only be achieved if 'the constituent farming/land-use decision makers within it have achieved ecological sustainability in the natural resource base; this in turn is dependent upon micro-economic sustainability or long-term economic viability of the individual farm units.' Finally, sustainability at the farm unit level depends upon agronomic sustainability or the ability of the farm's field system to maintain its level of production over a long period of time' (Smailes and Heathcote 1992).

(2) A further dimension, the question of social sustainability, needs to be added since the economic and ecological viability of a region is reflected in the social condition of its population. In the quest for economies of scale in agricultural production for a commercial market, the reduction of number of holdings is associated with a reduction of social contact and service provision which cannot be compensated for, despite Australia's innovative approaches to service provision over large areas (Royal Flying Doctor Service, School of the Air and Distance Education programs) and satellite television/communication systems.

(3) Politically there has been a long tradition in Australia of official support for the concept of the family farm, but recent studies have suggested that current circumstances are undermining the viability of the implicit small-scale and independent nature of such a production system (Lawrence 1987). This raises the question as to whether society wishes its agriculture to be managed by corporate or family-scale ownership and what the implications are of each for the sustainability considerations noted above?

(4) Finally, the research so far has illustrated the changing role of government in resource management on the Eyre Peninsula. Broadly, the trend has been from a role as 'real estate agent,' through that of 'developer's friend' and 'welfare agent' to that of 'steward of the nation's environment.' Associated with this trend has been a tendency for the increased centralization of official decision-making and a reduction of local control over local destinies. This tendency is probably global but will pose serious problems if sustainability joins the other policy issues in appearing to be imposed from the center upon the local peripheral societies.

REFERENCES

ABS 1986–7. Australian Bureau of Statistics, catalogue number indicated.

Berndt, R. M. 1985. Traditional aborigines. In *Natural History of the Eyre Peninsula*, ed. C. R. Twidale, M. J. Tyler and M. Davies, pp. 127–38. Adelaide: Royal Society of South Australia.

Boeree, R. M. 1963. Land settlement on Eyre Peninsula South Australia: an enquiry into the future of an essentially rural area based on the further development of its agricultural resources. PhD thesis in Geography, University of Adelaide.

Burgess, R. C., McTainsh, G. H. and Pitblado, J. R. 1989. An index of wind erosion in Australia. *Australian Geographical Studies* 27(1):98–110.

Burrows, K. 1989. Interpretation of South Australian rainfall data. In *Greenhouse '88: Planning for Climate Change*, ed. T. Dendy, pp. 34–9. Adelaide: Department of Environment and Planning.

Carder, A. 1981. Desertification in Australia: a muddled concept. *Search* 12:217–21.

Crocker, R. L. 1946. An introduction to the soils and vegetation of the Eyre Peninsula, South Australia. *Transactions of the Royal Society of South Australia* 70(2):83–107.

Ellis, S. and Porter, S. (n.d.) Administration of the 1976/77/78 drought in South Australia. Department of Agriculture, South Australia.

EPCSA. 1988. *The State of the Environment Report for South Australia*. Adelaide: Environmental Protection Council of South Australia, Department of Environment and Planning.

EPRCC. 1991. *Interim Report to the Minister of Agriculture*. Adelaide: Eyre Peninsula Regional Co-ordinating Committee.

Foley, J. C. 1957. Droughts in Australia; review of records from the earliest years of settlement to 1955. *Commonwealth Bureau of Meteorology Bulletin* 43.

Gentilli, J. 1971. Climates of Australia. *World Survey of Climatology* 13:35–211.

Gibbs, W. J. and Maher, J. V. 1967. Rainfall deciles as drought indicators. *Commonwealth Bureau of Meteorology Bulletin* 48.

Hare, K. 1987. Drought and desiccation, twin hazards of a variable climate. In *Planning for Droughts: Towards a Reduction of Societal Vulnerability*, ed. D. A. Wilhite et al. Boulder: Westview Press.

Heathcote, R. L. 1969. Drought in Australia: a problem of perception. *Geographical Review* 592:175–194.

Heathcote, R. L. 1988. Drought in Australia: still a problem of perception? *GeoJournal* 164:387–97.

Hill, R. 1987. Clear rainfall division between north and south. *The Advertiser*, Adelaide, 15 December.

Hobbs, J. E. 1988. Recent climatic change in Australia. In *Recent Climatic Change*, ed. S. Gregory. London: Belhaven Press.

Kimba District Council Minutes. Various dates. Kimba, South Australia.

Lange, R. T. and Lang, P. J. 1985. Vegetation. In *Natural History of the Eyre Peninsula*, ed. C. R. Twidale, M. J. Tyler and M. Davies, pp. 105–18. Adelaide: Royal Society of South Australia.

Laut, P., et al. 1977. *Environments of South Australia Province 4: Eyre and York Peninsulas*. Canberra: CSIRO Division of Land Use Research.

Lawrence, G. 1987. *Capitalism and the Countryside: The Rural Crisis in Australia*. Sydney: Pluto Press.

Lowrence, R., Hendrix, P. F. and Odum, E. P. 1986. An hierarchical approach to sustainable agriculture. *American Journal of Alternate*

Agriculture 14:169–73.

Meinig, D. W. 1962. *On the Margins of the Good Earth*. Chicago: Rand McNally.

Odingo, R. S. 1990. The definition of desertification and its programmatic consequences for UNEP and the international community. *Desertification Control Bulletin* 18:31–70.

Parker, A. J., Fanning, C. M. and Flint, R. B. 1985. Geology. In *Natural History of the Eyre Peninsula*, ed. C. R. Twidale, M. J. Tyler and M. Davies, pp. 21–46. Adelaide: Royal Society of South Australia.

Pittock, B. and Hennessy, K. 1991. Enhanced greenhouse climate scenarios. *Climate Change Newsletter* 32:4–5.

RDDEHCD 1978. *A Basis for Soil Conservation Policy in Australia*. Report No. 1. Canberra: Research Directorate, Department of Environment, Housing and Community Development.

Russell, J. S. 1981. Geographic variation in seasonal rainfall in Australia: an analysis of the 80-year period 1895–1974. *Journal of the Australian Institute of Agricultural Science* 47:59–66.

SAICVC 1976. *Vegetation Clearance in South Australia*. Adelaide: Report of the Interdepartmental Committee on Vegetation Clearance.

SAPP 1938. *Report of the Soil Conservation Committee*. Adelaide: Government Printer.

SAPP 1946. *Annual Report of the Department of Lands and Survey for 1946*. Adelaide: Government Printer.

Smailes, P. and Heathcote, R. L. 1992. Holding the line or chasing the rainbow? The quest for sustainability on Eyre Peninsula South Australia. *Progress in Rural Policy and Planning* 2:234–46.

Trewartha, G. T. 1964. Climates of the Earth. In *Goode's World Atlas*, ed. E. B. Espenshade Jr, pp. 8–9. Chicago: Rand McNally.

Trumble, H. C. 1948. Rainfall, evaporation and drought-frequency in South Australia. *Journal of Agriculture* 52:55–64 and Suppl. 1–15.

UN 1977. World Map of Desertification (1:25 000 000). Paris: FAO and UNESCO.

UN 1990. A working concept of desertification for the purpose of its assessment. *Desertification Control Bulletin* 18:28.

Wetherby, K. G., Davies, W. J. and Mathieson, W. E. 1983. Wind erosion on Eyre Peninsula 1975–1979. *South Australia Department of Agriculture Technical Report* 31.

Williams, M. 1974. *The Making of the South Australian Landscape*. London: Academic Press.

WPDSA 1982. Salting of non-irrigated land in Australia. Working Paper on dryland salting in Australia. Victoria: Soil Conservation Authority.

III

Climate Variability and Vulnerability: Causality and Response

6: Drought Follows the Plow: Cultivating Marginal Areas*

MICHAEL H. GLANTZ

Almost a century ago the belief that 'rain follows the plow' was a popular one that accelerated population movements into the region now known as the American Great Plains. Until the mid-1800s, this region was considered a wasteland, useless for agriculture and human settlement, an inhospitable obstacle to settlers in search of the promised land in the western part of the North American continent.

At the same time that settlements crept westward after the Civil War in the 1860s, rainfall became more prevalent. Suddenly, what had previously been considered a barren waste was seen as a potential garden and a breadbasket for the eastern part of the country. Many attributed the increased rainfall to the effects of human settling activities, which included plowing fields, creating ponds, irrigating dry areas, and planting trees on what essentially had been a treeless grassland. The railroad companies and land speculators seized on this explanation to convince Easterners to move to the Midwest. The advertising campaigns successfully produced waves of immigrants who came to the sparsely populated plains seeking their fortunes (Webb 1931).

During the 1890s a severe, multi-year drought dispelled the rain follows the plow theory. Thousands of settlers abandoned their homesteads to seek livelihoods elsewhere. It seemed that much of the support for the assumption of a causal relationship between rain and population was simply exaggeration by the railroads and other land speculators, intent on selling land at higher prices than it was worth. Before long, people realized that dry and wet periods commonly alternated and that the region's first surge of settlers had accidentally coincided with the onset of a lengthy wet spell. With the return of a prolonged drought, the credibility surrounding the rain follows the plow theory itself evaporated.

The scientific reasoning behind the belief is that a region's atmospheric circulation is positively affected by increased sources of evaporation. These derive from breaking the ground with the plow, creating open bodies of water (ponds and tanks) and planting trees whose roots suck the scarce moisture from the ground and whose leaves allow the water to evaporate into the atmosphere, which ultimately returns the moisture as rain.

Today the scientific literature is still filled with articles and studies on how land use has either brought about or eliminated rainfall in a region. The belief in a rain follows the plow concept still lives, although the number of its supporters appears to be small. The latest resurgence of interest in this idea resulted from the 17–year drought that began in 1968 in the West African Sahel. It was argued that the reverse process, the removal of vegetation, could create a desert-like environment (Charney *et al.* 1977). The overgrazing by livestock on vegetation and the collecting of firewood as the only available fuel ensured these conditions in the Sahel. Responding to the plow–rain theory, the United Nations and ALESCO (the Arab League Economic, Scientific and Cultural Organization) suggested that tree belts be built along the northern and southern edges of the Sahara Desert to arrest their encroachment, put moisture back into the air, and bring rainfall to desiccated areas.

It is my contention, however, that as agriculture moves into increasingly marginal areas, drought, not rain, is following the plow. Most of the world's best rainfed agricultural land is already in production or unavailable to production for political or other reasons. Increasing the amount of land for agricultural use requires the destruction of forests, the use of irrigation in arid areas, or the movement of people into marginal agricultural areas (Glantz 1994).

Recall that only 40 years ago, Nikita Khrushchev, then

* This article appeared in the April 1988 issue of *The World & I*, a publication of The Washington Times Corporation. © 1988. Reprinted with permission.

125

Secretary of the Communist Party of the former Soviet Union, launched his virgin lands scheme in an attempt to surpass United States grain production. The scheme's success would have demonstrated to the developing world that the Soviet Union was as much an agricultural force as an industrial power.

The plan 'encouraged' people to move into Soviet Central Asia and Kazakhstan and put arid and semi-arid land under mechanized agricultural production. Soon stories of the failure of this approach reached the press, as people began leaving the virgin areas to return home. Drought conditions plagued the virgin lands areas, and since that time rainfed agriculture has been supplanted by irrigation farming. The architects of the virgin lands strategy failed to take seriously the marginality of the climate for sustained production; they paid for it at the time, with humiliation in the economic development community. In fact, Leonid Brezhnev was intimately involved in the scheme's implementation and later wrote a book about the problems encountered in the virgin lands areas (Brezhnev 1978).

THE WEST AFRICAN SAHEL

The 1950s and 1960s were relatively wet across the West African Sahel. During this period the increased rainfall encouraged cultivators to move farther north toward the southern edge of the Sahara. The areas that they came to occupy, often with the encouragement of their governments, had traditionally been used as rangelands for livestock herds of the pastoralists. Annoyed and harassed by these new settlers, the pastoralists shifted their herding activities farther north. With the increased regional rainfall, more vegetation was now available farther north, so the pastoralists had what appeared to be equally good conditions. During this wet period, herds grew larger and required more rangelands and watering points. The new technology of drilling deep wells opened up even more land as seasonal pastures became usable with the availability of permanent sources of water (Monod 1975). As the wet period continued, more and more cultivators came to the region and began clearing the land for cultivation.

In 1968, an extended drought began, which some argue has lasted more than 17 years (Winstanley 1985). The first intense drought episode lasted from 1968 to 1973, claiming about 100 000 human lives and 12.5 million livestock. With the failure of the rains to move far north into the Sahelian zone, pastoralists and their herds became stranded at the deep wells. As a result, experts have suggested that most of the livestock that died during the drought perished from hunger, not from thirst.

When the drought wiped out their herds, many pastora-

lists went into refugee camps. Governments then sought to convert them to farming so that their contribution to the national economy could be encouraged and their production taxed. As pastoralists, they were perceived by their governments as independent, and thus a threat to central government. Moreover, they were hard to tax and only reluctantly participated in the modern sector of the economy.

It was in the West African Sahel that the image of the deserts on the march was first formulated. Speculation abounded that the region's climate was indeed changing. Rumors suggested that the desert's rate of advance was up to 50 kilometers a year. Closer investigation showed that the overgrazing of the southern edge of the Sahara during the wet period and the cultivation of areas deemed to be only marginally arable had created patches of desert-like conditions and caused sand dunes to form. With little vegetative cover, the dunes encroached on settlements.

In some of the regions where the government had encouraged cultivation, farmers planted grains and resorted to farming practices that did not take into consideration the long-term climatic conditions. When the drought recurred, as the region's rainfall history suggested it should, the farmers believed they were the victims of a natural disaster. Instead, they were victims of poor planning and a lack of understanding of the region's climatic record. After all, the 1968 drought was the third major one this century. Clearly, what had happened was that drought followed the plow in the West African Sahel (Glantz 1977).

BRAZIL

The plight of the inhabitants of the drought-plagued Brazilian Nordeste (Northeast) has inspired many classic Brazilian novels, such as da Cunha's *Rebellion in the Backlands*. Current patterns of land ownership and land use were established hundreds of years ago during Brazil's colonial period. Cattle ranching along the coast was supplanted by sugar cane production, forcing ranchers farther inland into the heart of the semiarid *sertão* (Magalhães and Magee 1994).

Land ownership in the Nordeste is a key factor in the ability of different segments of society to cope with severe drought. There are basically two types of ownership in the region: large estates and small subsistence landholdings. Estates are extremely large, sometimes encompassing a few hundred thousand square kilometers. In contrast, the poor are restricted to the marginal areas of this semi-arid region for subsistence food production and ranching. The scarcity of productive land is primarily a socioeconomic phenomenon: 60% of the population holds less than 10 hectares of land and 3% owns about 70% of the land (Hall 1978).

Subsistence farming takes place on small landholdings or on land leased to poor famers on the large estates. Even during years of good rainfall, the small farmer population is in danger because of low agricultural productivity. The growth of the population and of livestock herds in the region has forced farmers into increasingly marginal areas and has heightened the region's dependence on food produced on these lands. The soils and existing vegetation in these areas have thus become extremely vulnerable to climatic variability, explaining their persistently low and ever-declining levels of productivity.

The region is known to be plagued by recurrent short-term drought, as well as devastating multi-year dry spells. The poor farmers are the first to be hurt by drought, since they grow barely enough food for their own consumption. They must also find work to supplement their income, which they often do by working on the large estates as laborers. During extreme drought, however, the *latifundistas* (large land-owners) lay off their workers as a temporary cost-saving measure, making the workers' plight even worse. While the federal and state governments implement work programs to construct roads and dams for minimal wages, many men abandon their homes in search of work in other parts of Brazil.

The basic problem in the Nordeste appears to be one of land ownership: an expanding population trying to sustain itself on a dwindling resource base. More and more, we hear about the frequency and intensity of droughts, which is often correlated with the discussions of changing climate. The question remains open, however: Are we not witnessing the effects of a change in land use rather than in the climate, a situation in which drought once again has followed the plow?

AUSTRALIA

Australia has a long history of drought. In 1982–83 it witnessed its worst drought in more than a hundred years, with property and other losses estimated in the hundreds of millions of dollars. Brush-fires were rampant, consuming homes as well as vegetation. One might wonder why a developed country such as Australia is still so vulnerable to the impacts of drought.

Since the onset of European settlement in southeastern Australia, regional industry has shifted from sheep ranching to growing cereals and dairying. These early changes in land use, often driven by market demands and prices, began a process of climatic vulnerability, which advanced with the settling movement from the humid southern and eastern coasts toward the semi-arid and arid interior. In other words, human settlements were working their way down the rainfall gradients to regions where the climate was only barely

supportive. Population pressures mounted to put rangelands under the plow. As the wheat farmers moved down the rainfall gradient, they encountered increasing likelihood of drought (Heathcote 1994). The influx of farmers was eventually reversed, as many abandoned their land when the intermittent dry conditions, characteristic of the region, returned.

While government programs, which bought out farmers on the edge of bankruptcy, alleviated some of the pressures on the land, new technologies encouraged them to plant wheat. This reliance on wheat production further increased their vulnerability in a region known to be drought-prone (Heathcote 1986).

The severity of drought-induced crises in New South Wales can be attributed to the fact that farmers have moved into such a region and have thus put themselves at risk. The same is true for livestock owners who overstock their ranges and make themselves equally vulnerable.

A variable climate in that area caused prospective farmers to believe the climate had changed in favor of sustained agricultural or livestock production. Once drawn there, however, they were surprised by what they should have expected: a return to drier conditions. Yet, invariably, wet periods in the arid and semi-arid interior of New South Wales have lulled people into a false security, and when drought returns, farmers believe themselves to be the victims of a natural disaster.

CONCLUSION

The attempt to cultivate borderlands tends to marginalize both land and people. Most often, either a push or a pull into these areas initiates the process. 'Push' factors include population pressure, environmental degradation, and government policies. As populations continue to grow in developing countries and especially in sub-Saharan Africa, the natural per capita resource base decreases. If the best land is already in production, new farmers must move into previously farmed areas, which become further impoverished by overuse and misuse. Clearly, future generations will find it more and more difficult to support themselves from the land at all.

The 'pull' factors are relatively few. As mentioned earlier, climates fluctuate, sometimes with long wet or dry spells alternating. During wet periods in arid and semi-arid areas, the normally dry areas appear to be capable of sustaining agricultural production. Thus encouraged to move there, farmers often displace pastoral herders who have traditionally used these areas as rangelands.

The idea that drought follows the plow is based on the belief that increased pressures on currently used agricultural areas cause population movement into less productive, often marginal, areas. Consequently, we can expect to hear more in the future than we have in the past about droughts and their impacts on humankind (e.g. Glantz 1994).

REFERENCES

Brezhnev, L. 1978. *The Virgin Lands*. Moscow: Progress Publishers.

Charney, W. J., Quick, S. H. and Kornfeld, J. 1977. A comparative study of the effects of albedo change on drought in semiarid regions. *Journal of Atmospheric Science* **34**:1366–85.

Glantz, M. H. (ed.) 1977. *Politics of Natural Disaster: The Case of the Sahel Drought*. New York: Praeger.

Glantz, M. H. (ed.) 1994. *Drought Follows the Plow: Cultivating Marginal Areas*. Cambridge: Cambridge University Press.

Hall, A. 1978. *Drought and Irrigation in North-East Brazil*. Cambridge: Cambridge University Press.

Heathcote, R. L. 1986. Drought mitigation in Australia: reducing the losses but not removing the hazard. *Great Plains Quarterly* **6**:225–37.

Heathcote, R. L. 1994. Australia. In *Drought Follows the Plow: Cultivating Marginal Areas*, ed. M. H. Glantz, pp. 91–102. Cambridge: Cambridge University Press.

Magalhães, A. and Magee, P. 1994. The Brazilian Nordeste (Northeast). In *Drought Follows the Plow: Cultivating Marginal Areas*, ed. M. H. Glantz, pp. 59–76. Cambridge: Cambridge University Press.

Monod, T. (ed.) 1975. *Pastoralism in Tropical Africa*. London: Oxford University Press.

Webb, W. P. 1931. *The Great Plains*. New York: Grosset & Dunlap.

Winstanley, D. 1985. Africa in drought: A change of climate? *Weatherwise* **38**:75–81.

7: Amazonia and the Northeast: The Brazilian Tropics and Sustainable Development

JAN BITOUN, LEONARDO GUIMARÃES NETO and TANIA BACELAR DE ARAÚJO

INTRODUCTION

This paper examines the history of the relationship between Brazil's semi-arid Northeast and the tropical North, commonly known as the Amazon region.* The comparison provides a basis for a broader discussion of the possibilities for and constraints on sustainable development in the Amazon. The impetus for this study came from two sources. First, there is increasing international attention on development in the Amazon, and second, the Amazon and Northeast regions of tropical Brazil are becoming progressively more interdependent. Hence, it is time to question whether the scope of the present deliberations on the Amazon should be widened and placed within the context of the region's relations with neighboring states as well as the political economy of national development policy.

This study examines the roles played by the Northeast and the Amazon in the recent history of national development, and investigates the repercussions of changes in Brazilian society on people's perceptions of these geographic regions and of space in general. In particular, the chapter looks at the social and environmental impacts of changing development policies and associated demographics in the Amazon Region. The analysis focuses on the last 30 years with references to earlier periods.

In the first section of the chapter, we discuss the historical background of the perception that the two regions are peripheral to national development. Next the Northeast and the Amazon are placed within their national context, focusing particularly on the economic and social processes defining them as distinct regions. In the third section we explore the social and environmental impacts resulting from these distinctions. And in the final section we explore the significance of these historical interactions for the design of sustainable development policy today.

The history of the relationship between the Northeast, the Amazon, and other regions of Brazil

Modernization as Europeanization

The Northeast and the Amazon regions of Brazil are officially distinct political and administrative units. Until the early twentieth century, however, they were viewed by many Brazilians as a single region – indeed, they were grouped together as the vast wild 'North' in contrast to the smaller, more civilized 'South.' Although they share the common characteristic of being frontier territories, there are distinct differences between the two regions. These differences have grown in importance in recent decades.

This immense region did not undergo the systematic development which occurred in the Central West, the Southeast and the South of Brazil. The latter areas, where the State set guidelines for infrastructure development and modernization, were subject to intense European immigration. Instead, the 'North' was considered alien, in part because of its distinct non-white population, but also because of its tropical climate and forests.

Race has been an important defining characteristic of the Northern regions. According to the official commentator for the 1920 census, Oliveira Vianna, Brazil's future depended on the expansion of the white race – the only group capable of guaranteeing the ultimate victory of civilization over nature. In a study entitled 'Evolution of the Race,' appended to the 1920 census, he remarked that the process of 'arianization' in the North and Center was slow, but would ultimately be completed given 'greater fertility of the White element, compared to that of the Negro, the Indian, and the Mestizo'

---

* The Amazon region, sometimes referred to as the North or the Northern region, includes the states of Amazonas, Pará, Amapá, Rondônia, Acre, Roraima and Tocantins. The term Northeast refers to the states of Maranhão, Piauí, Ceará, Rio Grande do Norte, Paraíba, Pernambuco, Alagoas, Sergipe and Bahia.

He proclaimed that 'The population that emerges from the racial melting pot to which the Northeast was being subjected, shall be an Aryan dressed in our tropical attire; the same being the case with the rest of the country' (Oliveira Vianna 1922:344). Vianna ascribed to the Blacks and Indians what he called an 'ethnic chaos' which inhibited progress. Hence, with respect to the formation of a Brazilian national identity, the elites considered the historical 'North' as inferior, in large part due to its undesirable concentration of populations of non-European descent. With time, this racial distinction was overcome.

Nature and development of the North

In addition to its ethnic characteristics, the 'North' was also distinguished by its natural features. The 1920 census divided the Brazilian territory into three climatic zones: (I) subtropical (the region between the tenth parallel and the Tropic of Capricorn); (II) temperate (south of the Tropic of Capricorn); and (III) tropical (north of the tenth parallel).

The tropical zone included the Northeastern states located north of the state of Bahia, and the states and territories of Amazonia. Both regions were difficult to settle. From the point of view of the state, the main problem was how to settle populations permanently in tropical settings considered naturally hostile to man. Since the Great Drought of 1877-9 in the Northeastern *sertão* (or 'hinterlands'), large outmigrations have placed pressure on neighboring regions. The endless forests of the Amazon Basin were considered a logical destination for Northeastern outmigrants.

In practice, however, the settlers were unable to transform the Amazonian environment into the stable and productive area they sought. While describing the flow of Northeastern colonizers moving into Acre in the Amazon region, Oliveira Vianna wrote:

> The Great Drought of 1877 caused the rugged cowhands from Ceará to begin invading the Amazonian forests. Each fierce climatic cycle casts into these solitary forests crowds of colonizers bearing prodigious physical resistance. The new environment in which they settle is entirely different from that of their homeland: they trade a dry, warm and healthy climate and come to settle in a humid, unhealthy one; a landscape of bushes and xerophytes for one of forests and water and overwhelming vastness. The intrepid people conquer this harsh land, establish communities and turn Amazonia, for a brief moment, into a center of wealth as active and important as the coffee-producing cultures established by the incomparable effort of the farmers of the South ...
>
> There is, however, in the midst of this rapidly growing society, an inevitable instability inherent in the nature of their economic base – rubber tapping. This settler is not there to become rooted to that land, but rather as a

transitory exploiter of the land's wealth, moving on as the yields of the rubber trees cease. (Oliveira Vianna 1922:306)

In short, retaining as well as Europeanizing populations within these 'hostile' regions was of great concern to the state from early on.

Development and regionalization

As the capacity of the state expanded and technology improved, a more precise knowledge of the Brazilian territory emerged. When the authoritarian Getulio Vargas government assumed power in 1937, it created the Instituto Brasileiro de Geografia e Estatistica (the Brazilian Institute for Geography and Statistics, IBGE) to organize economic planning data and activities. The first major activity conducted by the IBGE in the late 1930s and the 1940s sought to develop a stable statistical data base and cartographic survey for use by the government to help identify the national territorial potential. The second major initiative undertaken by IBGE, at the end of the 1960s, was designed to provide guidelines for regional development policies and national integration through the creation of a unified market.

In 'The regional repartition of Brazil' Guimarães proposed that the concept of natural regions be used to divide Brazil for statistical and didactical purposes. This method relied on natural resource surveys to identify the defining characteristics of each region (Guimarães 1941). Guimarães suggested that the semi-arid climate be a decisive factor in delimiting the Northeast, and that the rain forests delimit the North, or Amazon region. In reality, this may not be the most accurate method for establishing regions as, for example, transitional rain forests are present in the western part of the Northeast, specifically in the states of Piauí and Maranhão. Nevertheless, this characterization defined the primary development problems of each region. In the Northeast, the lack of water associated with the expanding population and increased pressure on the land was viewed as the principal cause of poverty. In the Amazon region, the lack of transportation and communication infrastructure was considered the major restriction on the exploitation of 'inexhaustible' natural resources.

Subsequently, the state identified differences between the Northern regions and the South other than simply physical geography, and a new approach to regional definition and development was sought. Studies of the national urban network during the early 1970s classified the Northeast and the Amazon as 'peripheral zones,' in contrast to the more developed 'Center,' hence, industrialization and the faster distribution of goods determined the new contours of Brazilian geography. Urban transportation networks were envisioned as the vector for this diffusion of innovation and manufactured commodities (Faissol 1973).

Faissol, in his article entitled 'The Brazilian urban system,' placed the central development nucleus around São Paulo and relegated the relatively underdeveloped areas of the interior of the Northeast, the Great Amazonian Region and the Center-West to the periphery. Faissol was concerned with the problems of national integration. 'The structure of the sub-system in peripheral areas is characterized by low rates of urbanization and the diffuse power of mid-sized and small cities; this was the target of the urban development policy – to establish relations of greater reciprocity between the two great areas [of northern and southern Brazil]' (Faissol 1973).

The Northeast and Amazon regions continued to be regarded as an aggregate in spite of sharp differences in the formation of their social and physical settings. A policy that considered the Northeast and the North as a single unit could not possibly take into account the complex social and environmental conditions of these regions. It is in examining their differences, however, that the characterization of the Northeastern and the Amazon regions and regional policies must be sought.

The emerging regional vision and development institution
The Northeast was defined and integrated into the national identity much earlier than the Amazon. Freyre, in his 'Regional Manifesto,' and subsequent works, clearly discussed the basis of Northeastern social formation and its importance in defining Brazilian identity (Freyre 1937). The geographer, de Andrade, observed in his text *The Land and Man in the Northeast* that it was the organizational rules of the great sugar-producing plantations of the Northeast that decisively characterized land use and social relations in the region (Andrade 1963). In the late 1950s, the rural social movement for agrarian reform shifted the debate on regional problems from a theoretical to a political concern. In 1959 the Superintendencia do Desenvolvimento do Nordeste (Superintendency for the Development of the Northeast, SUDENE) was created to direct development in the region by orienting investments which supported national goals with fiscal incentives.

The Superintendencia para o Desenvolvimento da Amazonia (Superintendency for the Development of Amazonia, SUDAM), the parallel organization for Amazonia, was founded 10 years later. Urban development, which was considered strategically important for national economic development, was to be facilitated by these organizations. Industrial poles were created in the larger cities of the Northeast and a Free Trade Zone was established in Manaus, aimed at making that city an urban pole for Western Amazonia.

The emphasis on urban development integrated the previous perceptions of each region. In the Northeast, irrigation projects in the Great São Francisco Valley were of prime concern; in the Amazon region, infrastructure construction was considered the key to opening the new 'frontier,' where people, lured by easy access to land, would become the labor force with which to exploit the unlimited natural resources. In the 1970 National Integration Plan, the two regions were viewed as complementary: the Northeast would supply labor and the Amazon region would supply land and natural resources. The linkage was made between Amazonian development and the poverty problems of the Northeast (Panagides and Magalhães 1974).

THE NORTHEAST AND THE AMAZON REGION WITHIN THE NATIONAL ECONOMY

Recent economic processes

Although the Northeast and the Amazon Regions have been regarded as a single region in Brazilian history, they have played different roles. Beginning in the sixteenth century, at the zenith of colonial Brazil, the Northeast was the first region to experience economic development. This development was centered on the sugarcane industry, which still plays an important role in the economies of the Northeastern states of Alagoas, Pernambuco, Paraíba and Rio Grande do Norte. The Amazon region developed later with the advent of rubber tapping, which prospered in the late nineteenth century. The rubber boom lasted less than half a century (1870 to 1920), but prompted the first surge of migration from the Northeast to the Amazon. During that period, Manaus, the capital of the province of Amazonas, grew from a village of 300 to a 'metropolis' of 75 000 people, becoming one of the ten largest cities in Brazil (Lacerda de Melo and Moura 1990:33). The flow of population between the two regions, which began at this time, still persists.

In the late 1950s, Brazil sought to integrate into the world economy by becoming a major world power. Under the slogan 'Brasil Grande Potencia' (Brazil Great Power), the government initiated major industrialization and development efforts, launching Brazil into the category of a newly industrialized country (NIC). This economic and technological surge was accomplished with significant support from the multinational and international financial community. The 20 years between 1955 and 1975 were marked by great economic expansion and the intense modernization of Brazil. The Brazilian government took an active role in investments, subsidies, market protection and regulation. This development was accompanied by rapid urbanization as well as greater concentration of wealth and power.

During this rapid economic transformation, the various regions became increasingly interdependent, and local and regional economies were integrated, although regional productive structures remained relatively distinct. Thus, the Northeast and Amazon regions have been affected in recent years by the overall changes imposed by urban-industrial economic consolidation, which is dominated by oligopolistic firms. A study of the Brazilian national economy in recent decades concluded that regional development in Brazil has been defined by the industrialization projects developed in Brazil after 1955 and the oligopolistic economic concentration characteristic of this development (FUNDAJ 1990).

Guimarães Neto's regional economic analysis identified four processes, listed below, that have had significant effects on economic life in each of the Brazilian regions (Guimarães Neto 1990a:131–164).

1. *Market expansion*
In the 1950s and 1960s, industries based in São Paulo broadened their commercial base by seeking markets in other regions. These regions were unable to accumulate enough local capital to foster any inter-regional competition with São Paulo's industry. The Brazilian government intervened, whenever possible, to stimulate the modernization of local industries. For example, SUDENE successfully stimulated the traditional textile industry of the Northeast. By the end of the 1950s, however, regional industries were faced with a dilemma: modernize and 'de-regionalize,' or become extinct. The North experienced a similar situation: the industrialized urban centers of the region and their peripheral markets were saturated by industrial production from Rio de Janeiro and São Paulo.

2. *Economic integration*
The integration of production, which predominated in the 1970s, was based on the regionalization of the large public and private oligopolies. This process did not eliminate the broadening of the commercial base, but rather overlapped and even reinforced the flow of commodities among the various Brazilian regions. The State boosted the productive integration of these regions through fiscal and financial incentives, particularly in the case of the Amazon and the Northeast (Araújo 1982).

With the flow of large capital investments, the less-industrialized regions experienced important changes. First, typical capitalist relations of production were introduced into the regions. Second, because the growth and modernization of the economies of the North, the Northeast and the Center-West were based on outside capital which exceeded the resources available to local agents, these areas experienced rapid and intense change. Finally, the dynamic economies of the so-called 'peripheral regions' were tied, during expansion and crisis, to the overall dynamism of the Brazilian economy which was dominated by the Southeast. Therefore, instead of

the 'traditional' model of regional economies dominated by agrarian export production, Brazil developed into a 'regionally localized national economy' (Oliveira 1977), whose growth, stagnation or recession affected multiple regions in unified waves.

3. *Physical and territorial integration*
The importance of physical and territorial integration for the relationship between the North and the Northeast will be discussed in detail below. It is worth noting, however, that the desire for consolidation of the internal market in conjunction with strategic military interests, propelled the development of the national transportation and communication systems linking remote regions of the country. This integration made possible the opening of resource frontiers and intensified migration, with significant consequences in the North and the Northeast.

4. *Inclusion in the world economy*
Brazil's determination to be included in the world economy, characteristic of the period known as 'the Brazilian economic miracle', was in part a result of the need to generate surpluses to service the external debt. Public enterprises and the government actively encouraged exports, reinforcing the 'regionalization' of capital, prompted first by the internal market growth. Evidence of this regional process includes the modernization and re-orientation of agriculture in the state of Paraná; expansion of grain agriculture within the Center-West; and the exploitation of mineral resources in the North, particularly in the state of Pará, and in the Northeastern state of Maranhão.

In conclusion, the processes described above and Brazil's dynamic economy changed the regional productive structures of both the North and the Northeast, as well as other regions of Brazil. The modernization and consolidation of industry expanded tertiary activities which in turn increased the rate of urbanization and created successive waves of migration from the Northeast and South into the Amazon Region, as illustrated in Table 1.

Economic decentralization

Both early twentieth century industrialization and the development of heavy industry in 1955–75 were focused in the Southeast. During the recent cycle of economic expansion (1955–75), and the subsequent economic slow-down and crisis, there were strong counter-tendencies to this geographic concentration. The Northeast and the Amazon were participants in this decentralization process.

IBGE figures show that, between 1970 and 1985, the Southeastern share of the Gross Domestic Product dec-

Table 1. *Brazil, the Northeast and the Amazon: population and production ratios, 1960–85 (%)*

	Brazil				Northeast				Amazon			
	1960	1970	1980	1985	1960	1970	1980	1985	1960	1970	1980	1985
Agricultural product/total GDP	20.7	11.2	9.7	10.1	24.6	18.7	16.1	—	40.2	19.4	16.3	16.2
Industrial product/total GDP	27.7	30.4	37.2	35.5	17.0	15.3	37.2	—	14.4	15.1	30.3	30.1
Tertiary product/total GDP	51.6	58.4	50.6	57.5	58.4	66.0	46.7	—	45.4	65.4	53.4	53.4
Urban population/total population	44.7	56.0	67.6	72.0	37.4	45.1	51.7	53.5	33.9	41.8	50.5	45.4
Rural population/total population	55.3	44.0	32.4	28.0	62.6	54.9	48.3	46.5	66.1	58.2	49.5	54.6
Urban EAP/total EAP	—	57.7	70.1	74.1	—	43.0	56.8	—	—	37.5	50.1	56.4

Note: GDP, gross domestic product; EAP, economically active population.

Sources: FGV/IBGE, National Accounts (up to 1980) for the regions and every year for Brazil; SUDENE for the Northeast since 1985; IBGE, Demographic Census; PNAB 1985; Annual Statistics for 1987.

reased from 65% to 58%, while increases were registered in the Amazon (2.2% to 4.3%), the Northeast (12% to 13.5%) and the Center-West (3.7% to 6.3%). The South maintained its share at about 17%. Geographic concentration of industrial activity is still an important aspect of the Brazilian economy, however. An IBGE study, based on 1980 census figures (IBGE 1990), showed that only 12 urban centers in Brazil have an Industrial Value Added (IVA) tax that exceeds 1% of the national total. Together, these centers comprise nearly two-thirds of the country's industry, with the São Paulo metropolitan region alone generating one-third of the national IVA. Among these 12 centers, two are located in the Northeast (Salvador and Recife) and one in the Amazon (Manaus) (Fig. 1).

Examples of recent decentralization in these regions include:

(1) Establishment of a chemical industry zone in the Northeast, between Salvador and Maceió. This zone includes Bahia's petrochemical growth pole, the integrated base complex in Sergipe, and the chloro-chemical pole in Alagoas. In 1980, Salvador accounted for 12% of the national chemical industry's IVA, exceeded only by São Paulo which accounted for 21%.

(2) Establishment, in Manaus, of an important pole for the manufacture of durable consumer goods. With 13% of the national output of electric and communication materials, Manaus ranks second only to the São Paulo metropolitan region (which produces 51%).

(3) Development of Recife's plastics products industry (3.7% of total IVA), ranking fifth in the country. For electric and communications materials, Recife accounts for 1.7% of the national IVA, ranking eighth.

(4) Establishment of the mining and metallurgical complex at Carajás in the Amazon, with another center in the Northeast, in the state of Maranhão.

(5) Beverage production in Recife (fourth-ranked producer holding 4% of the national IVA) and Salvador, in the Northeast, and in Belém in the Amazon (sixth-ranked producer with 3.2% of the IVA).

(6) The mining industry in the Amazon (Amazonas and Pará states) and the growing timber industry in the region.

For decades the Southeast has been losing its relative importance in agriculture and livestock production. In the 1950s, the South accounted for 55% of the country's output of agriculture and livestock. In the 1980s, however, this share fell to one-third of the Brazilian total. By contrast, the South, the Central-West and the Amazon region have become progressively more important producers, together accounting for nearly half of the national output. The Northeast's share of the national output in agriculture and livestock has not changed significantly. Agricultural output in the Amazon has recently increased, however, due to the incorporation of new agricultural areas in the Amazon Basin and in the Cerrado. In the 1970s alone, nearly 71 million hectares of new agricultural area were brought into production. These areas were located mainly in the Central-West and in the north and west of the states of Maranhão and Bahia.

Finally, it is worth noting that these trends in the decentralization of industry may not continue in the near future (Cano 1989). In the manufacturing sector, for instance, new technologies aimed at saving labor, energy and conventional raw materials (especially in the steel, aluminum, paper, and cellulose industries) may affect the Amazon and the Northeast, whose comparative advantage has been cheap labor. On the other hand, it is likely that new sectors, such as information technology and micro-electronics, will concentrate in the Southeast where the physical infrastructure and level of education are considerably higher than in the Amazon and the Northeast.

In summary, the Amazon and the Northeast have experienced the following trends in recent years:

Figure 1: Regions and states of Brazil.

(1) *Growth, diversification, and modernization of their productive base*; and increases in their share of the country's economic output from 14% to 18% of total product.

(2) *An inflow of public investments and private capital* from national and foreign sources (Fig. 2). In the Amazon, these investments were directed at the extractive and manufacturing industries. The Northeast received investment in these areas, as well as in transportation and storage services.

(3) *Greater 'outward' integration* of their regional spaces, as a result of strong efforts to consolidate the national internal Brazilian market. These efforts have increased consumption of consumer goods. Integration has also been strengthened by the 'regionalization' of large-scale capital investments. Figures available for inter-regional trade in 1980 show, for

instance, that the Northeast exported mainly to the Southeast and South, which bought 83% and 10%, respectively, of goods produced in the Northeast. Of all commodities imported by the Northeast, 85% originated in the Southeast and 11% in the South. By comparison only 5.4% of the Northeast's output was exported to the Amazon, and this region supplied only 2.2% of Northeastern imports (SUDENE 1985).

As illustrated, the Northeast had a surplus domestic trade balance with the Amazon. In 1980, the region sold 48% more than it purchased from its neighbor. Northeast exports to the Amazon included: plastic, tobacco, sugar, machinery and electrical equipment, mechanical instruments, clothing and textile accessories. Northeast imports from this region

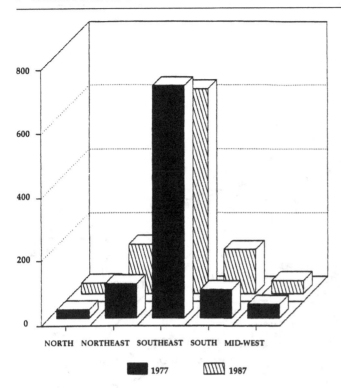

Figure 2: Location of the 1000 largest companies in Brazil, 1977 and 1987.

these two regions. The development of energy generation operations in the Amazon, primarily for the purposes of industrial consumption in the Northeast, constitutes another example of the linkage between these two areas.

In the 1960s and 1970s, small farmers from the Amazon and the Northeast shared in the production of grains, especially low-quality rice. The southern parts of Pará and Maranhão ranked fifth in national rice production. According to the Superintendencia Nacional do Abastecimento (National Supply Superintendency, SUNAB), this rice was destined for consumption by the low-income strata of both regions; but it also reached the markets in the Southeast (according to Cibrazem's figures, it provided 10% of Rio de Janeiro's consumption), as cited by Guilherme Velho (1972:123).

Migration, facilitated in large part by improvements in transportation, has been the most intense and significant linkage between the Amazon and the Northeast. The causes and consequences of this migration will be analyzed in the following sections.

Physical and territorial integration

The process of migration between the Northeast and the Amazon preceded the development of a viable road network. Improved infrastructure, however, has secured the link between the two regions and has increased the volume and frequency of these migrations. In addition, migration has also tended to concentrate in particular areas of the Amazon, exerting more pressure on them (Fig. 3).

Migration from the Northeast to the Amazon region was historically facilitated by two 'natural' means of transportation. Seafarers arriving in Belém proceeded upstream and dispersed throughout the Amazon river basin in search of latex and other extractive forest products. This migratory pattern played a major role in the deforestation and agricultural exploitation of the region of Bragantina, located to the east of Belém. Further to the east, in Maranhão, colonies of settlers formed within the forested areas along rivers such as the Mearim and the Pindaré. Within Maranhão, Andrade identified a second migration route established by cattle ranchers heading south in the direction of the *cerrados* (a sparse type of dryland vegetation), between the Parnaíba and Tocantins rivers. Large ranches were established along the trails linking Floriano (Piauí) and Carolina (Maranhão) (Andrade 1977).

Reaching the Tocantins, this 'cattle front' became the second link between the Northeast and the Amazon. It drew migrants who settled throughout the forested areas north and west of Carolina. The opening of a trail which followed the telegraph line between the Pindaré and Tocantins basins

included: wood and wooden products, tobacco, machinery and electronic equipment, textile articles, ceramic products, vegetable textile fibers and cereals. The most dynamic development poles in the Northeast are increasingly integrated to the Southeast's economy, however. Examples include the chloro-chemical and petro-chemical complexes, the metal-mechanical industries, the clothing and textile operations, the mining-metallurgical complexes, and agro-industrial and grain operations (Duarte 1990). The strong backward linkage of the new intermediate products industries to the natural resource base in the Northeast is important to highlight (Souza 1986:60–3).

In the Amazon, the agricultural, cattle-ranching, mining, and extractive forestry industries likewise are anchored in the regional resource base. Nevertheless, the modern and dynamic production of durable consumption goods, localized in Manaus, demonstrates a very fragile articulation, regarding the creation of income and the generation of employment, with the local economic base.

(4) *Economic integration among states and the development of complementary activities.* Examples of these cases are: the mining-metallurgical complex of Carajás which encompasses the states of Pará and Maranhão, the expansion of large subsidized agricultural and cattle-ranching projects (located in the western part of the Northeast and in the eastern part of the Amazon), and the growth of timber exploitation, the furniture industry and construction within

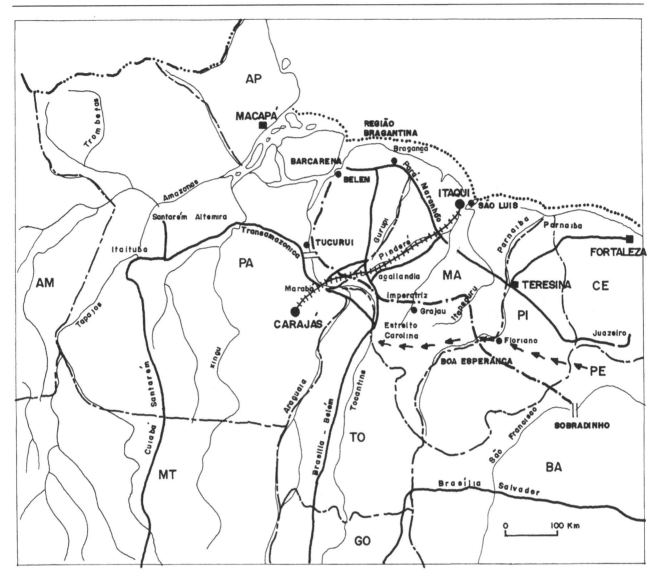

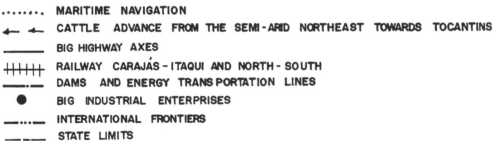

Figure 3: The physicoterritorial integration of Amazonia and the Northeast.

(linking the cities of Grajaú and Imperatriz) strengthened this migratory trend.

The more recent changes in migration patterns, and in the physical and territorial integration of these two regions, were caused by the construction of major highways under the 1944 National Highway Plan. This plan was designed in response to strategic concerns regarding national security and was implemented in the Amazon under the military government. The newly opened highway corridors directed and concentrated the flow of migrants, beginning in the 1960s. In addition, the construction projects attracted laborers, many of whom remained in the region after the conclusion of the projects.

In this context, it is worth noting the roles played by the

Pará–Maranhão and the Belém–Brasília highways. The Pará–Maranhão road linked the Bragantina region of Pará to the Pindaré valley in Maranhão, thereby replacing the sea route between these two states. The Amazonian segment of the Belém–Brasília highway offers an alternative route to the old river axis of the Tocantins, which linked Estreito to the outskirts of Belém. The Transamazonian highway, implemented by the National Integration Plan of the early 1970s, constituted, in its northeastern segment (from Cabedelo to Estreito), the second link between the two regions. In stretching as far westward as Altamira and Itaituba, it transformed the region around Estreito–Imperatriz into a base for the settlement of the Belém–Brasília highway (paved in 1973) and provided further consolidation of the region with the opening of the Santa Inês–Açailândia highway, paved during the 1980s.

Migration was also fostered by large hydro projects aimed at harnessing the power of the Amazon basin rivers and supporting the mining resources of Carajás. These projects drew mostly on labor from the middle Tocantins, as well as labor moving along the Belém–Brasília axis. Major projects in the Amazon region include the Tucuruí dam and the transmission lines linking Boa Esperança (CHESF), Tucuruí (ELETRONORTE) and Barcarena–Belém; iron mining in Carajás; and the construction of the Carajás–Itaqui/São Luis do Maranhão railway (via Açailândia).

The large influx of migrants has given rise to frequent disputes regarding access to land in the area, which have often turned violent. The pressures bringing migrants into the Amazon region have resulted in a range of social, economic and environmental problems, many of which raise politically sensitive issues. This process was epitomized by the gold panning at Serra Pelada; now that Serra Pelada is closed, other areas are creating the same problems.

The development of large highways and large public works projects has progressively transformed the Eastern Amazon into a region where Northeastern migrants are providing labor and settling alongside the highways, under the power transmission lines, and around large state-owned projects. This physical and territorial integration of the region has not, however, been met with commensurate economic and social development. As mentioned previously, the two regions developed a greater outward integration rather than an inward one. The Belém–Brasília highway, the North–South railway project and the export-oriented nature of the Itaqui and Barcarena ports, contribute more to national and international linkages than inter-regional relations between the Northeast and the Amazon.

Inter-regional relations and public investment in infrastructure have made migration to the Amazon from the Northeast and the South possible. They have also created a territory from which resources of national and international value have been drained. Access to the tremendous resources of the Amazon and the use of Northeastern labor for their exploitation have not generally benefited the population of these two regions despite the integration of these regions into the national and global economy.

Patterns of migration

The intense migration to the Amazon from the Northeast in recent years is only the most recent manifestation of a longstanding pattern. Furtado, in his *Formação Economica do Brasil* (Economic Formation of Brazil), related migration to labor history, and compared the flow of people from the Northeast to the Amazon in the late 1800s to the simultaneous patterns of migration from Europe to Brazil. The intense demand for rubber, associated with the production of the internal combustion engine, focused attention on the economic potential of the Amazonian region and boosted the demand for labor for rubber extraction and processing. Rubber prices increased ten-fold and generated a significant increase in production from 6 thousand tons during the 1870s, to 11 thousand in the 1880s, and to 35 thousand tons in the first decade of the twentieth century. The author estimated that, between the beginning of the boom and the first decades of the twentieth century, approximately 500 000 people from the Northeast migrated to the Amazon.

The reservoir of labor existing in the interior of the Northeast proved invaluable in meeting this intensive labor demand in the Amazon. This migration was not only in response to an intense demand for labor. The prolonged 1877–80 drought, which destroyed almost all of the cattle and subsistence agriculture, was, as noted by Furtado, cleverly utilized to direct migrants, who were offered organized support services, subsidies for transportation and propaganda to attract them into the Amazon. Even at the peak of the demand for labor, however, the condition of the migrants, who generally arrived indebted and indentured, can only be considered a form of serfdom. During the drought crisis, misery was generalized and the migrant's life was degraded to the most primitive subsistence conditions (Furtado 1977).

Since that time, the factors determining migration to the Amazon have been, on the one hand, the physical and territorial integration of the region with the rest of Brazil and the expansion of the agricultural and industrial frontiers, and, on the other hand, recurring drought and chronic underdevelopment that actively pushed people into the region.

Origins of migration to the Amazon
A superficial analysis shows that migration from several regions in Brazil toward the Amazon is relatively small in

relation to the flows moving to regions such as the Southeast. These flows, however, are growing steadily with each 10-year census period. There were 487 200 Northeast immigrants living in the Amazon in 1980; in the Southeast, which has been the traditional destination of immigrants from the Northeast, the number in the same year was 4.3 million, out of which 2.9 million alone resided in the state of São Paulo.

It is estimated that migration to the Amazon and its frontier grew significantly during the 1980s as a result of the recession and deepening social crisis (the so-called 'lost decade') that hit Brazil, particularly the urban economy of the Southeast. Unfortunately, 1991 census figures are not yet available to show the extent of this flow. Migration toward the Southeast has declined and many of the migrants in this direction are ending up in the Amazon frontier region. Figures from the Ministry of Labor used as indicators for formal urban employment show that the employment levels stagnated during the period 1979–88 in São Paulo and Rio de Janeiro. Rio experienced an employment decline of 2.3% during this period while São Paulo experienced a negligible increase of 6.8% (Guimarães Neto 1990b).

Preliminary figures from the Agricultural Census of 1985, when compared with the 1980 census figures, show that the expansion of rural employment and of small agricultural establishments were still significant. The number of people employed in the Amazon increased from 1.78 million in 1980 to 2.23 million in 1985, a yearly increase of 4.6%. Between 1970 and 1980 the increase averaged 6.2%. The number of rural establishments smaller than 100 hectares increased from 336 300 to 414 000 between 1980 and 1985, a yearly increase of 4.2%. The area occupied by these establishments, however, increased from 7.3 million to 9.9 million hectares, a yearly increase of 6.3%.

The preliminary figures from the 1991 Demographic Census show significant population increases in the Amazon, in contrast to a decrease in the national population growth rate from 2.5% for the period 1970–80, to 1.9% for 1980–91. The population of the state of Rondônia, for instance, increased at a yearly rate of 7.9% between 1980 and 1991, Roraima 9.6%, Amazonas 3.5%, Pará 3.7%, Amapá 4.7%, and Acre 3% (Table 2).

There are several complex processes that explain the migration patterns to the Amazon. Migration from the Northeast can be explained by a number of factors which include drought, the land-tenure structure, the changes in economic activity which resulted in reduction in the rural labor demand, and the need for employment outside the region.

During the 1970s, three long periods of drought had significant impact on production and population of the semi-arid (*sertão*) Northeast (Fig. 4). The drought in 1970 forced 500 000 people to join public works projects; drought was

Table 2. *Amazon region: recent population growth, 1980–1*

	Population (1000)		Annual rate (%)
	1980	1991	
Northern region	5 800	9 226	4
Rondônia	491	1 130	8
Acre	301	417	3
Amazonas	1 430	2 089	4
Roraima	79	216	10
Pará	3 403	5 084	4
Amapá	175	298	5
Brazil	119 003	146 154	2

Source: IBGE, Demographic Census.

repeated in 1976 and in the early 1980s (1979–83). The latter forced the enlistment of 2.7 million people at government work projects, representing nearly one-third of the economically active rural population of the semi-arid zones. This triple climate-induced disruption of the rural economy within a decade, which affected half the region's population (17.9 million people), surely accelerated outmigration from both the rural and urban areas of the Northeast (Carvalho 1988).

The second critical factor contributing to underdevelopment and the relatively low labor absorption in the Northeast is the pattern of land ownership. It is important to consider questions related to land use not in isolation but in conjunction with the concentration of capital and economic power brought about by the structure of land tenure in the region (Andrade 1963; Figueroa 1977; Carvalho 1988; Pessoa 1990). The lack of access to official financial resources, the low levels of education, the absence of effective associations or cooperatives to strengthen their bargaining power, the barriers to access to land, and the general weakness of civil society keep the small farmer in a permanently tenuous condition, particularly in the face of adverse climatic conditions and the existing structure of production in the region. This prevailing structure of concentrated ownership and economic power stifles opportunity for change and provides the large landowner, the merchant and the financial agent with conditions to appropriate significant shares of small-farm production. The proletarianization of the sugarcane workers and the spread of cattle ranching activities in areas traditionally under subsistence agriculture precipitated the expulsion of workers from these rural zones (Lacerda de Melo 1976).

The construction of the major highways facilitated the exodus from the Northeast. Beginning in the 1950s, and continuing throughout the height of highway construction in

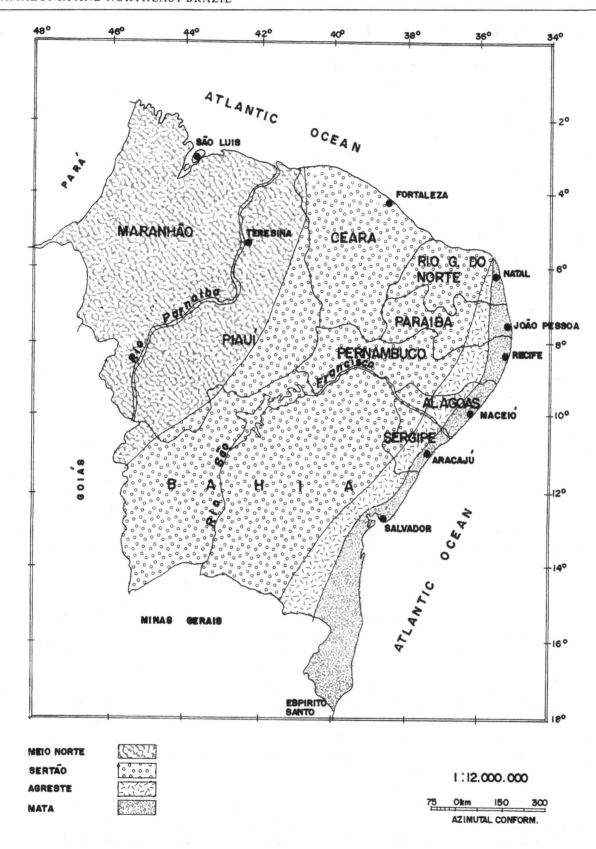

MEIO NORTE

SERTÃO

AGRESTE

MATA

1 : 12.000.000

75   0km   150   300

AZIMUTAL CONFORM.

SOURCE   ANDRADE , MANUEL CORREIA DE A TERRA E O HOMEM NO NORDESTE

Figure 4: Geographic zones of the Northeast of Brazil.

the 1970s, the settling of the transverse axis of the Transamazonian highway and of the Perimetral Norte, and the intra-regional axis such as the Cuiabá–Santarém and Porto Velho–Manaus highways, was a key objective of the National Integration Program (PIN) (Becker 1990).

The accelerated transformation of other Brazilian regions, particularly the South and the Southeast, further affected migration patterns from the Northeast. The 1970s 'economic miracle,' and the concentration of industrial activities in the South, brought about radical transformations in productive processes and in labor relations, especially in rural areas. In conjunction with these changes came a further concentration of land ownership, and the substitution of labor-intensive crops for capital-intensive crops such as soya. As a result, the rural exodus was accelerated and small farmers who could not afford to modernize, migrated to the urban centers of the South and the Southeast, and the dynamic rural zones of the Central-West and the Amazon (Martine and Camargo 1984; Taschner and Bógus 1986).

Despite the growth of migration to the Amazon from the Brazilian South, Northeasterners constituted the most numerous group emigrating to the Amazon. Moura and Santos (1990) studied the intensity of emigration from the micro-regions of the Northeast during the 1970s. The areas of highest intensity are: (1) the eastern frontier of Maranhão; (2) the state of Ceará and the micro-regions of Campo Maior and Valença in Piauí, and Paraíba (Sertão da Cajazeira and Depressao do Alto Piranha); (3) 'in the part of the *agreste* (an intermediary type of ecological region found between *sertão* and the tropical coastal forests) located in the state of Pernambuco ... extending as far as the state of Alagoas,' and (4) the southern part of the interior of Bahia (Moura and Santos 1990). Those Northeasterners who headed to the Amazon went first to the state of Pará and secondly to Rondônia. Immigrants originating in the South and Southeast generally went to Rondônia or Pará. Those from the Center-West followed a similar pattern to that of the Northeasterners.

There were three major migrations to the Amazon: one from the Northeast to the state of Pará (73% of the total); one from the South and Southeast to the state of Rondônia (58% from the Southeast and 74% from the South), and one from the Center-West to the state of Pará (62%). Pará was the most attractive destination, receiving 56% of this total, followed by Rondônia with 33%. Preliminary data for the 1980s indicate that Rondônia emerged as the preferred destination in that period (Fig. 5).

Maranhão and Ceará contributed more than two-thirds of the emigrants arriving in the Amazon region during recent decades. Maranhão had until recently itself been a destination of intense migration originating in the semi-arid zones, where the concentrated land-owning structure of the North-

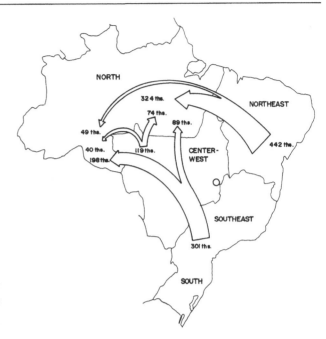

Figure 5: Migration flows for the northern region of Brazil, 1990. ths., thousand. (Source: Historical Statistics of Brazil, IBGE 1987.)

east was reproduced and further contributed to the expulsion of people toward the Amazon.

In summary, the factors that precipitated migration to the 'frontier' of the Amazon were (1) the opening and consolidation of the highway network; (2) failure to develop and improve living conditions in the semi-arid regions of the Northeast, and to offset the devastating impact of the recurring droughts; (3) the inability to solve the problems of agrarian organization and concentration of land ownership; and (4) the inability of urban centers in the Northeast and other regions to absorb the growing labor force which included those who were expelled from the rural zones and those who already lived in urban centers who could not find employment in industry or services. In brief, the persistent vulnerability to climatic variations and chronic underdevelopment of the Northeast region, together with the overall deterioration of Brazil's economic situation, made the Amazon an attractive destination to those 'expelled' from the Northeast and other regions.

THE SOCIAL AND ENVIRONMENTAL IMPACTS OF THE INTEGRATION OF THE NORTHEAST AND THE AMAZON

In earlier sections of this chapter, we argued that the industrialization of the Brazilian economy and consolidation of a more unified domestic market brought about the regional integration of Brazil. This transformation reinforced the links among the Amazon, Northeast, the rest of the country, and the world economy. It resulted, as well, in the development of economic complexes and large enterprises in the Amazon and Northeast. The physical and territorial integration brought about by highway construction and the implementation of large development projects in the Amazon and Northeast, facilitated migration into the Amazon. This migration had considerable impact on social and environmental conditions in both regions; the following sections of this study will examine these impacts.

It is not possible to isolate the social and environmental impacts related to any single emigration episode. Nonetheless, by looking at the destinations chosen by a particular group of emigrants and the prevailing type of occupation found within the new communities, some important inferences will be drawn out here about who migrates from the Northeast, their living conditions, and their relationship with the Amazonian physical and biological environment.

Social impacts

Some preliminary considerations must be taken into account before a specific analysis of the social impact of migration to the Amazon from the Northeast can be made. First, migration in past decades took place in what were considered demographically empty spaces. In recent decades, this has not been the case. Regardless of their origin, migrants to the Amazon will confront a set of established interests, including a pattern of land ownership and concentration of wealth even more accentuated than in the places they left behind. This process has important implications for the kind of labor relations and opportunities migrants face. According to the 1985 Agricultural Census in the Amazon region in 1980, 577 rural establishments larger than 5000 hectares (0.1% of the total number in the region) covered 31% of the total area occupied by all regional establishments (nearly 14 million hectares). Twenty-two establishments larger than 100 000 hectares totalled 6 million hectares. In contrast, 229 600 establishments smaller than 20 hectares, 46% of the agricultural establishments in the region, occupied a mere 3.3% of the total area surveyed by the census.

To a large extent this extremely unequal distribution of land is a consequence of the inability of the Brazilian State to exert even minimal control over the direction and organization of settlements in the Amazon. This lack of state management of migration in the region has led to growing conflicts over land in the states of Maranhão, Pará, Mato Grosso and Rondônia, involving 82 000 families and an area of about 5.7 million hectares (Becker 1990:40). At the heart of these conflicts lies the appropriation of land by agricultural and cattle ranching enterprises, farmers, Indians, tenants, colonizers and former farmers, together with the agents of the large landowners, speculators and opportunists. These conflicts are over Indian lands, unoccupied government-owned lands, and lands destined for official settlement programs. Timber interests seeking to access lands belonging to Indians, tenants, settlers and rubber tappers have also aggravated these conflicts (Becker 1990:41, 42).

Agriculture and cattle ranching in the region, in the context of an extremely unequal agrarian structure, relies on the appropriation of land rather than on improvements in techniques to increase productivity (Kageyama 1986:181). The capital-poor migrants from the Northeast – the great majority of settlers – sought to establish small family production units, either by accessing land directly on the expanding Amazonian frontier, or through a landowner. Most migrants also work for wages. Jobs are usually temporary and may absorb small farmers who seek them as a complementary source of income. The wage market, however, consists of a recruitment chain involving agents (known as *gatos* or *gateiros*) who act as team leaders and price their middleman services at half the wages of each worker. These wage jobs usually consist of field mowing for individual farms and companies, and rice harvesting.

In a 1990 study, Becker found that the majority of migrants in the Eastern Amazon were from the rural Northeast. She pointed out that though the mobility of the Northeast migrants allowed a more efficient use of capital, from the point of view of the migrant such migration generally implied instability and exploitation. In her words, mobility creates 'difficulties in organizing political action and trade unions. Nevertheless, mobility also represents a survival strategy of the peasant family, a process of social consciousness and aspirations' (Becker 1990:51). It is unquestionably a painful path to social awareness.

Employment and income conditions in the Amazon region are not significantly better than those in the Northeast. After making adjustments for the underemployed, about one-third of the rural labor force in the region was estimated to be unemployed in 1980 – one of the highest rates in the country. In the same year, average income of the region's economically active population was 1.38 times the minimum wage, while the figure for the Northeast was 0.79 times the minimum wage. The average income in the Amazon between 1970 and 1980 was one the lowest in the country; only the Northeast was lower. Forty-nine percent of the workers employed in cattle ranching and agriculture received wages equal to or below the minimum wage (Kageyama 1986). According to Kageyama, the poorest 0.5% of the economically active rural population in the Amazon region experienced a decrease in their share of the total income from 30.2% in 1970, to 23.85% in 1980. This decrease occurred despite significant expansion in the region's agricultural production during this same period.

In addition to agricultural and ranching activities, many of the migrants to the Amazon turned to gold panning. Until the 1960s, panning for gold was carried out by approximately 10 000 workers. Today, Becker estimates nearly 240 000 workers are panning in the region – 80% of the active panners in the country. Gold panning is subject to some of the same exploitative relations and conflicts that are characteristic of the agricultural sector. Intense disputes involving territorial and mining claims take place not only between miners and state-owned, national and multi-national companies, Indian nations and gold panners, but among gold panners as well.

Some points must be highlighted with reference to gold-mining migrants and their condition: (1) the majority of gold panners are migrants from the Northeast; (2) workers receive either daily allowances or a minimum percentage share of the boss's income (normally the boss is the representative of the owner of the mine); (3) the worker's income is hardly sufficient to meet basic needs for subsistence, mainly due to the fact that the worker depends on suppliers who appropriate a significant part of the economic surpluses generated in their activity.

The itinerant nature of small-scale mining, involving constant relocation of workers, is another source of conflict over resources in the Amazon region. Conflicts between the panners and the mining companies are compounded by conflicts with the Indian nations. Currently, nearly 14% of the Amazon region has been granted as government concessions for prospecting and mining activities. Five hundred and sixty of these concessions are located within Indian territories. The 1700 gold prospecting concessions pending include 33% of the total Indian areas in the region, particularly those in Pará and Rondônia (Becker 1990:79).

While settlement in rural areas is still the dominant pattern of Amazonian migration, the process of urbanization has intensified. As the agrarian structure is consolidated, and the control over land and natural resources, including minerals, is strengthened, migrants are left with fewer and fewer opportunities for creating a livelihood. Migration to the urban centers becomes one of the few alternatives left to them. The crisis and stagnation of the national economy in the 1980s had the same socioeconomic impact on the urban economy in the Amazon as in the rest of the country: increased underemployment and declining incomes.

We may conclude that, thus far, the living conditions of migrants to the Amazon, both urban and rural, are better than in the Northeast. Migration, however, has resulted in the exchange of one marginal survival strategy for another. For most families, their first residence is but the beginning of a long journey. The itinerant character of much of the agricultural and mining activity demonstrates this phenom-enon. The pervasive instability of settlements is perpetuated by the changing economic bases of the regional urban centers which depend on easily disrupted and depleted resources such as panning, agriculture, cattle ranching or large development projects (state-owned or not).

Environmental impacts caused by the occupation of the Amazon

In this section we identify some of the causes and consequences of environmental degradation in the Amazon region of Brazil, in particular in eastern Amazonia where most of the migrants from the Northeast have concentrated. We try first to identify those factors which affected the development of settlement policies in the region. We then relate these policies to the environmental degradation taking place, and analyze how they have given rise to the national and international controversy about the destruction of the rain forest.

The first factor that needs to be understood is the official perception of the Amazon region as a simple supplier of space and resources for Brazilian development. Characteristic of this perception was the National Integration Plan of 1970, whose motto was 'a land without people for people without land.' The soil was regarded as an inexhaustible resource capable of sustaining growing populations in the vast Amazon, especially to those from the Northeast who were in need of better land and water. The Amazonian soils, however, have not assured success for the agricultural settlements.

Two to three percent of the total land area of the Amazon is considered good quality agricultural land. The soils known as 'terra firme' (poorly drained lands) represent the majority of the Amazonian soils. Gourou wrote: 'The major portion of the Amazonian terra firme soils are unquestionably mediocre, though not necessarily useless for agricultural purposes. In spite of their apparent uniformity, these soils present a certain variety which should influence their use' (Gourou 1982:193). The agronomist Falesi illustrated the mediocrity of the soils and their variability in a detailed map called 'Edaphic Environment of the Great Carajás Project' – an area of 895 265 square kilometers comprising 40% of the state of Pará, 95% of the state of Maranhão and the northern area of the state of Tocantins. He states that half of the area is comprised of highly leached, nutrient-poor, lateritic (7%) and podzolic soils (12.34%) lying on varied topography ranging from flat to moderately sloped. These soil conditions result either from the paleoclimatic history of the region or from the prehistoric and historic deforestation. Less than 10% of the area is covered by fertile, humus-rich soils (Falesi 1989).

These recent studies reveal an Amazonia that is for the most part mediocre for agriculture but still largely unknown, concealed under an exuberant forest cover. The traditional uses of these *terra firme* environments were modest and extensive, oriented to extractive production rather than to settled intensive agriculture. The shifting cultivation system of indigenous peoples, based on manioc and supplemented by fishing, hunting, forest resources extraction and rubber tapping, was dispersed throughout the vastness of the forest. One region has been moderately successful at adopting a productive agriculture, however: the Bragantina region and its northern extensions located on the southern part of the Guamá river. In this region deforestation was very intense due to the consumption market for firewood, charcoal and manioc flour in Belém. In the forested areas located to the south of the strip stretching from Belém to Bragança, which exhibits relatively high rural densities, the cultivation of *malva*, an antiseptic medicinal plant, has had some success (Sawyer 1979); there have also been some thriving black pepper plantations that use techniques resembling those adopted by the Japanese colonies in Tomé-Açu (Monteiro 1987).

The random pattern of recent occupation, the lack of prior familiarity with the complexity of the soil distribution, the lack of interest in evaluating the modest experiences mentioned above, as well as the tendency of the migrant from the Northeast to equate exuberant forest cover with good soils, have all led to the degradation of the areas of *terra firme* without any economic return or, at least, any of a sustainable nature. Environmental deterioration is particularly evident along highways which have been built invariably between riverbeds to avoid costly engineering constructions. These highways have been the vectors of the occupation of the *terra firme* environments along the Pará–Maranhão, Belém–Brasília, Transamazonian, Cuiabá–Porto Velho, Cuiabá–Santarém, and Porto Velho–Manaus routes.

The government programs for settling rural workers in *terra firme* areas, whether undertaken with or without the National Integration Plan (PIN) guidelines, have all resulted in failure. The reasons for this failure are: impossibility of keeping the colonizers concentrated in the agro-villages which, in spite of being equipped with a minimum of necessary services, were distant from their plots of land; difficulties in marketing products due to the distances to the commercial centers and poor road conditions; rapid impoverishment of the soils once the forest cover was burned; as well as the presence of malaria, which reduced the working capacity of the settlers.

Following the construction of the roads, there was a spontaneous settlement of small farmers who had been progressively driven from better lands by the expansion of large landholdings in the region. One concentration of small farmers was confined to an area suggestively called *Trecho Seco* (Dry Stretch) which was entirely inappropriate for small-scale agriculture. The confinement of these farmers to such areas along the Belém–Brasília highway has forced them to search for alternative means of survival by panning, working at construction camps, and migrating to cities (Bitoun 1980). The undefined legal status of the lands prevented successful settlement by these small farmers, which could only have been achieved if these lands had easy access to markets and were located in the rare areas with good or moderate agricultural potential.

By the early and mid 1970s, it was widely recognized that this form of resettlement as a model of development would not be successful. This gave rise to an exploration of alternative models for the *terra firme* areas. One such model has been the attempt to promote colonization through enterprises, i.e. allocating plots and colonizers in a similar initiative as that in the north of Pará, where coffee production had been expanded. The region of Alta Floresta (in Mato Grosso) is also an example of this model. To what extent have these enterprises effectively taken a technical and commercial lead? Or have they limited their actions simply to negotiating the plots or taking advantage of cheap labor in order to prepare their pastures? The answers given to each case entail success as well as failure, in settlements located in areas distant from the consumption markets.

Throughout these development efforts, soil continued to be regarded as a resource capable of generating significant surpluses for the national and international markets. Hence, rice production in areas recently settled in small family units has achieved temporary significance in supplying city markets outside the region. Production has progressively declined, however, due to losses in soil fertility and the disruption of the small family units by the expansion of large properties.

The state initiated financing and credit schemes for large cattle-ranching enterprises with the idea of transforming the Amazon into a major meat-producing center for national and international markets. Large ranch holdings began expanding in 1973, succeeding the deforestation and soil disruption caused by the small rice growers. The majority of these large projects have not proven to be economically viable, as a result of: sanitation problems; effects of the seasonal droughts – which become progressively more intense with distance from the equator; and impoverishment of the pasture soils which then require capital intensive management which is uneconomic. According to Gourou: 'Instead of chopping down large forested areas, it would have been wiser to cultivate small tracts using intensive techniques and thus provide forage for semi-confined livestock' (Gourou 1982:231). This author suggests the use of soya and manioc (with its leaves and roots), and regards semi-confinement as a means of better monitoring animal health. Obviously, this form of cattle ranching would supply regional markets rather than transform the Amazon into a meat exporting center.

Although soils of the *terra firme* terrains proved difficult to manage, the wealth of their forest cover rewarded those who, after the rubber tappers and the owners of nut-producing areas, have appropriated them. The accumulated experience of the traditional groups engaged in forest activities has only recently drawn the public interest: their knowledge of the species has not been utilized as yet. The fact that *Bertholletia excelsa* produces autotoxins that prevent concentration of the species in small areas has long been known. But how much of such knowledge has been applied to practical silvicultural attempts? Small enterprises and small-farm communities can invest in biodiversity, with the assistance of traditional knowledge. Examples of attempts to increase local biodiversity through cultivation include the introduction of fruit and cocoa cultivation in Tomé-Açu, where black pepper previously constituted the sole product. There have also been promising attempts, in the municipality of São Joao do Araguaia, at cultivating the *cupuaçu* (a local vegetable) in association with the maintenance of the nut trees (Monteiro 1987; Souza 1991). Despite this new awareness of the productiveness of minor forest products, the predatory exploitation of the forest assets of the Amazon by the timber and cattle industries continues.

Lower wetlands have been particularly vulnerable. They occupy only a small part of the Amazon – about 12%, of which only 2% are considered fertile (Falesi 1989). Gourou estimates that wetlands cover 60 000 of the 3.5 million square kilometers of the Brazilian Amazon (Gourou 1982:196). A large proportion of the fertile alluvial soils are not accessible, however, as they are either located near small riverbeds or are susceptible to floods whose control would require enormous dams. The majority of the soils of the wetland forest (*igapó*) are poor and the wealth of the flora derives mostly from the ecological exchanges conducted with the trees of the flooded forest. Once more, the vitality of the vegetation conceals the modesty of the valuation potentials. It is in these wetlands, however, that the most varied types of living creatures are found. While in the *terra firme* areas economic exploitation has been extensive, the traditional productive activities in the wetlands, although 'primitive and rugged' have illustrated a striking human adaptation to the limits of the environment (Lacerda de Melo and Moura 1990:151).

Historically, the wetland areas of the Amazon region have given us the best insight into sustainable development practices. The outstanding character of land management in these areas lies in the complementarities among aquatic, forest and agricultural environments. Fishing provides the fundamental nourishment for the rural and urban populations. The forests provide game and fruits for consumption or trade, such as the *açaí* (a fruit used for its juice), as well as latex from the rubber trees. The flood plain soils are exploited according to a complex agricultural calendar and local micro-zonation, producing manioc, banana, fruits and vegetables, as well as industrial crops such as jute and *malva*. As

the distances to the consumer markets decrease, cattle ranching on the flood plain grasslands increases in significance (Gentil 1983).

These production systems have been inadequately studied, not least for their value as a 'conservation' of techniques for the exploitation of the Amazonian environment. The perception that traditional methods of land use are outdated and stagnant has inhibited greater investment in wetlands. On the contrary, they merit significantly more study, for they could become the basis for a more rational, although perhaps less ambitious, development of the wetland areas. In fact, the forms of occupation developed in these areas since the 1960s have totally ignored environmental rationality: the great hydroelectric projects drown these environments, while panning activities on the riversides ruin their fishing wealth by discharging mercury. The long-range destructiveness of these two forms of resource use must be included in any management accounting, just as the localized and limited nature of wetlands is considered.

In conclusion, the limited attempt to reconcile traditional and modern land uses in wetland areas as well as in the vast *terra firme* areas, reinforces the opinion expressed by Mário Lacerda de Melo in his writings:

> In the face of wildly optimistic assessments of the conditions and natural resources of Amazonia, voiced in the sixteenth century by Pero Vaz de Caminha in Brazil, and later by Humboldt, or on the other hand, exceedingly pessimistic views questioning the viability of the whole vast region, there is one position that can be perceived as realistic. Such a realistic position acknowledges, in the vastness of Amazonia, the existence of severely limiting factors but, at the same time, recognizes that the environmental adversities can be progressively overcome, through broad-based and persistent efforts (Lacerda de Melo and Moura 1990:123–4)

Lacerda de Melo's position is supported by Gourou, who wrote:

> The effort and capital spent in the economic development of Amazonia have not been in vain, even if they have not produced the expected returns. There has been a dash of megalomania in some projects. The destiny of the Indians and the Brazilians in Amazonia has not been given the deserved attention. The large cattle ranches, perhaps, are not a good way of utilizing Amazonia. The errors can be corrected: instead of dreaming about millions and millions of hectares we should be realistic in carefully assessing the value of the tens of thousands of hectares where tiny and prudent clearings will be opened in the Amazonian jungle. (Gourou 1982:238–9)

CONCLUSIONS

In brief, the analysis shows that, in recent decades, a model of development that concentrates wealth and excludes certain social groups has prevailed in the Northeast and in the Amazon. These two regions, regarded as peripheral in this process, have endured and participated in an economic and social evolution characterized by its predatory nature in relation to both the human population and the natural environment. Though this common experience has served to unify these areas, the resulting social and economic transformations have brought few improvements in the living conditions of their populations. Furthermore, the economic processes taking place have failed to acknowledge the distinct environmental reality of the semi-arid Northeast and the Amazon.

It has also become evident that recent Brazilian development efforts, following an industrialization and urbanization model, have favored certain areas of Brazil. In particular, center–periphery relationships have been established that have imposed conditions on the Northeast and the Amazon which have not always met the specific interests of their populations. The efforts at decentralization in the past decades have not been effective and have reinforced the peripheral character of these two regions. Additionally, in cases where sub-regions have participated more actively in the economic development and industrialization process, this participation has linked them more closely with the extra-regional economy than with the pre-existing regional productive base.

Recent economic development, particularly the 1950s, 1960s and 1970s, has not addressed the agrarian problem, nor has it encouraged a search for new forms of organization of the agricultural economy. In the Northeast and the Amazon, where large portions of the population are dependent on agriculture, these questions are of utmost importance. Economic development policies have failed to address the concentration of agrarian landholdings, the exploitation of rural labor, or the appropriation of the economic surplus generated by farmers or landless by large landowners or other elites. Without a direct attack on the existing production and labor relations, which have their roots in the concentration of rural property, a whole range of intermediate relations are maintained by mercantile capital in order to appropriate the income of the unassisted and non-organized farmer. This has denied communities the resources that could enable them to improve their living conditions and develop their productive capacity.

The agrarian structure described is rooted in the economic, social and political history of the Northeast and reproduced in the Amazon, sometimes in a more dramatic fashion. There, the degree of land concentration is greater and the conflicts, in the absence of a strong state, have tended to become generalized – involving immigrants, large landowners, speculators, indigenous peoples, lumber companies, small miners, and the Government through large, state-owned enterprises.

Sustainable development in this context has been difficult to achieve. Neither in the semi-arid tropics nor in the humid tropics have we been able to bring about development which respects the environment while systematically providing improvements in the living conditions of local populations. This is precisely the focus of the most recent debates on economic development, which have given rise to the exploration of sustainable development. As a consequence of the national and international concern regarding the future of the rain forests, proposals that address sustainability issues in the Amazon have been gaining strength. These proposals will not be viable if political and inter-regional considerations are not analyzed, along with economic and technological criteria.

Broad preconceptions shape interpretations of the development needs of the Amazon and Northeast. In the case of the Amazon, it is the vision of a vast open land and a resource frontier in need of a communication network. In the Northeast, the limiting factor is water, and the solution is seen to lie in the development of irrigation. New proposals, however, must address the social and environmental heterogeneity of the two regions. While the integration between these two regions and other areas has increased, a number of differences between the regions exist. These differences have not been adequately addressed by those presenting new development proposals.

The present study challenges us to apply the world's technological advances to support rather than supplant the knowledge and experience of local populations. This combination of knowledge and experience constitutes an invaluable asset that can generate more productive solutions, compatible with the existing social and natural environments. In particular, the wetlands areas in the Amazon and those of dry agriculture in the Northeast should be objects of systematic and long-term research by those studying sustainable development, as they are areas where human populations have developed a long-term productive interaction with nature.

In the political arena we must attack outdated agrarian structures and the exploitative nature of production and labor relations. We must also make the prevailing commercial and financial systems accountable. These structures and systems prevent farmers from appropriating a significant part of the economic surplus generated by their own work, and results in the impoverishment of millions of people (World Bank 1990:67,74).

Finally, it is clear that any proposal for sustainable development for the Amazon and the Northeast must take into account the important inter-regional links between these two areas. We have attempted to demonstrate that the economic links between the semi-arid tropics and the humid tropics have increased in the last decades. It is clear now that these links have prompted the development of new areas such as the Eastern Amazon, where labor from the semi-arid Northeast is used to produce commodities that are shipped predominantly to foreign markets and to the Southeast of Brazil.

Expansion of the transportation and communication infrastructure in the region has supported migration from the Northeast to the Amazon, and the transport of goods from each of these regions to the rest of Brazil. Models aimed at relieving the demographic and social pressure on semi-arid areas through the occupation of tropical humid regions have resulted in negative social and environmental impacts in the latter regions, and have not solved the problems of the semi-arid areas, particularly their vulnerability to recurrent droughts.

We therefore conclude that sustainable development policies for the semi-arid areas of the Northeast that do not respect the environmental diversity and the accumulated local experience, and do not seek to change long-standing structural agrarian inequities, will continue to increase the pressure on the Amazon's space and resources. Because of the dramatic increase in accessibility between the two tropics over the last decades, any policy aimed at sustainable development in the Amazon is doomed to failure unless a complementary proposal is adopted in the Brazilian semi-arid tropic of the Northeast.

ACKNOWLEDGMENTS

This study was sponsored by the National Council for Scientific and Technological Development (CNPq). Translation from Portuguese was done by Norma Gerjoy.

REFERENCES

Andrade, M. C. de 1963. *A Terra e o Homem no Nordeste*. São Paulo: Brasiliense.
Andrade, M. C. de 1977. *Paisagens e Problemas do Brasil*. São Paulo: Brasiliense.
Araújo, T. B. de 1982. Estado e industrialização do Nordeste. *Revista Economia e Desenvolvimento*. Rio de Janeiro: Campus.
Becker, B. K. 1990. *Amazonia–São Paulo*. São Paulo: Atica.
Bitoun, J. 1980. *Ville et développement régional dans une région pionnière du Brésil: Imperatriz–Maranhão*. Thesis, University of Paris.
Cano, W. 1989. *Reestruturação Internacional e Repercussões Inter-regionais nos Países Subdesenvolvidos*. Reflexões sobre o Caso Brasileiro. Campinas. (Mimeo.)
Carvalho, O. de 1988. *A Economia Política do Nordeste (Seca, Irrigação e Desenvolvimento)*. Rio de Janeiro: Campus.
Duarte, R. 1990. Dinâmica e transformação da economia Nordestina na década de 70 e anos 80. *O desenvolvimento desigual da economia no espaço brasileiro*. FUNDAJ. Recife. (Mimeo.)
Faissol, S. 1973. O sistema urbano Brasileiro: uma análise e interpretação para fins de planejamento. *Revista Brasileira de Geografia* 35(4):3–34.
Falesi, I. C. 1989. O ambiente edáfico da região do programa Grande Carajás. *Revista Brasileira de Geografia* 50(4):7–29.
Figueroa, M. 1977. *O Problema Agrário no Nordeste do Brasil*. Recife: HUCITEC/SUDENE.
Freyre, G. 1937. *Nordeste: Aspectos da Influência da Cana Sobre a Vida e a Paisagem do Nordeste do Brasil*. Rio de Janeiro: Published by the author.
FUNDAJ/INPSO/Departamento de Economia. 1990. *O desenvolvimento desigual da economia no espaço brasileiro: década de 70 e anos 80*. Recife. (Relatório de Pesquisa, mimeo.)
Furtado, Al-o. 1977. *Formação Econômica do Brasil*. São Pãolo: Ed. Nacional.
Gentil, J.M.L. 1983. A juta na agricultura de várzea da área de Santarém. Dissertation, Federal University of Pernambuco.
Gourou, P. 1982. *Terres de Bonne Espérance*. Paris: Le Monde Tropical Plon.
Guilherme Velho, O. 1972. *Frentes de Expansão e Estrutura Agrária*. Rio de Janeiro: Ed. Zahar.
Guimarães, F.M.S. 1941. Divisão Regional do Brasil. *Revista Brasileira de Geografia* 3(2):318–73.
Guimarães Neto, L. 1990a. Questão regional no Brasil: reflexões sobre processos recentes. *Cadernos de Estudos Sociais* 6(1):8
Guimarães Neto, L. 1990b. Mercado de trabalho na década perdida. *São Paulo em Perspectiva* 4(3/4):131.
IBGE 1990. *Brasil: Uma Visão Geográfica dos Anos 80*. Rio de Janeiro: IBGE.
Kageyama, A. 1986. *Modernização, Produtividade e Emprego na Agricultura. Uma Análise Regional*. IE/UNICAMP.
Lacerda de Melo, M. 1976. *Proletarização e Emigração nas Regiões Canavieiras e Agrestinas*. Departamento de Ciências do Homem da UFPE. Recife. (Mimeo.)
Lacerda de Melo, M. and Moura, H. A. de 1990. *Migrações para Manaus*. FUNDAJ. Recife: Massangana.
Martine, G. and Camargo, L. 1984. Crescimento e distribuição da população Brasileira: tendências recentes. *Revista Brasileira de Estudos de População*. 1(1/2).
Monteiro, A. A. A. 1987. *Permanência da pequena produção em Tomé-Açu: da pimenta do reino ao cacau*. Dissertation, Federal University of Pernambuco.
Moura, H. A. de 1972. Variações Migratórias do Nordeste 1940/1970. *Revista Econômica do Nordeste*. 14(Oct–Dec):35.
Moura, H. A. de and Santos, T. de F. (eds.) 1990. *Proteção da População do Nordeste por Microregiões 1980–2000*. FUNDAJ. Recife: Massangana.
Oliveira, F. de 1977. *Elegia para uma Re(li)gião*. Rio de Janeiro: Paz e Terra.
Oliveira Vianna, F. J. 1922. *Recenseamento do Brasil*, vol. 1. Ministério da Agricultura, Industria e Comércio. Diretoria Geral de Estatística. Rio de Janeiro: Typ. da Estatística.
Panagides, S. and Magalhães, V.L. 1974. *Amazon Economic Policy and Prospects*. Florida: University of Florida Press.
Pessoa, D. 1990. *Espaço Rural e Pobreza no Nordeste*. FUNDAJ. Recife: Massangana.
Sawyer, D. 1979. Colonização da Amazônia: Migração de nordestinos para uma frente agrícola no Pará. *Revista Econômica do Nordeste* 10(3):773–812.
Souza, A. do V. 1986. *Política de Industrialização, Emprego e Integração Regional*. O Caso do Nordeste do Brasil. Dissertation, CME/PIMES/UFPE, Recife. (Mimeo.)
Souza, C. H. L. 1991. *A Produção do Espaço e a Organização do Território em Ubá e Araras, PA*. Dissertation, Federal University of Pernambuco.
SUDENE. 1985. *Nordeste: Exportações e Importações, 1974–1980*. Recife.
Taschner, S. P. and Bógus, L. M. M. 1986. Mobilidade espacial da população Brasileira: aspectos e tendências. *Revista Brasileira de Estudos de População*. 3(2).
Tocantins, L. 1961. *O Rio Comanda a Vida*. Rio de Janeiro. Civilização Brasileira.
World Bank. 1990. *World Development Report: Poverty*, French edn. New York: Oxford University Press.

8: Reducing the Impacts of Drought: Progress Toward Risk Management

DONALD A. WILHITE

INTRODUCTION

Images of malnutrition, famine and a degraded African landscape were commonplace during the 1980s and appear likely to continue well into the 1990s and beyond. Glantz (1987) has shown that drought has hindered the ability of much of sub-Saharan Africa to achieve a sustained level of agricultural production and, as a result, has retarded progress toward economic development. Linkages between drought and economic development, although most obvious in Africa, exist throughout much of the developing world.

The impacts of drought in developed countries differ substantially from those experienced in much of the developing world. Absent are the widespread occurrences of food shortages, which may lead to malnutrition and famine, and large-scale evidence of land degradation. However, economic costs, particularly in the agricultural, energy and transportation sectors, are substantial. The recent droughts in the United States and Canada have been stark reminders of the vulnerability of all nations to this extreme climatic event. This increased awareness of the economic, social and environmental costs of drought is leading a growing number of nations, both developed and developing, to seek a more proactive approach to drought management. These nations now realize that they can no longer afford to divert scarce financial resources to drought relief programs that do little to reduce, and may actually increase, vulnerability to subsequent periods of water shortage.

The purpose of this chapter is to present a case study of the recent progress that has been made by some countries in reducing their vulnerability to drought. Progress toward drought preparedness will be discussed for the United States of America, South Africa, and Australia. Although many countries have made significant progress in this area in recent years, these countries were chosen because of what is considered dramatic philosophical changes in the way drought

and its management are perceived by government. The intent is for other drought-prone regions to examine these approaches and consider adapting them to their particular social, political and environmental setting.

The chapter is divided into three parts. The first will present an overview of drought and drought planning that will serve as background information for later discussions of recent drought policy and program changes in the United States, Australia and South Africa, all discussed in the second section. The final section provides a brief description of a generic planning process that is being promoted as one method of developing comprehensive preparedness plans for dealing with future episodes of drought.

DROUGHT OVERVIEW

Drought as a natural hazard

Drought differs from other natural hazards (such as floods, hurricanes and earthquakes) in several ways. First, since the effects of drought accumulate slowly over a considerable period of time, and may linger for years after the termination of the event, a drought's onset and end are difficult to determine. Because of this, drought is often referred to as a 'creeping phenomenon.' Second, the absence of a precise and universally accepted definition of drought adds to the confusion about whether or not a drought exists and, if it does, its severity. Third, drought impacts are less obvious and are spread over a larger geographical area than is damage that results from other natural hazards. Drought seldom results in structural damage. For these reasons the quantification of impacts and the provision of disaster relief are far more difficult tasks for drought than they are for other natural hazards.

Although drought represents a considerable climatic risk in semi-arid regions, it is a normal part of the climate for virtually all climatic regimes. Drought differs from aridity since the latter is restricted to low-rainfall regions and is a permanent feature of climate. The character of drought is distinctly regional, reflecting unique meteorological, hydrological and socioeconomic characteristics. Many people associate the occurrence of drought with the Great Plains of North America, Africa's Sahelian region, India or Australia; they may have difficulty visualizing drought in Southeast Asia, Brazil, Western Europe or the eastern United States, regions perceived by many to have a surplus of water.

Drought should be considered relative to some long-term average condition of balance between precipitation and evapotranspiration in a particular area, a condition often perceived as 'normal' (Wilhite and Glantz 1985). It is the consequence of a natural reduction in the amount of precipitation received over an extended period of time, usually a season or more in length, although other climatic factors (such as high temperatures, high winds and low relative humidity) are often associated with it in many regions of the world and can significantly aggravate the severity of the event. Drought is also related to the timing (i.e. principal season of occurrence, delays in the start of the rainy season, occurrence of rains in relation to principal crop growth stages) and the effectiveness of the rains (i.e. rainfall intensity, number of rainfall events).

Responding to drought: a historical perspective

Governments have traditionally relied on a wide range of potential actions to deal with the impacts of water shortages on people and various economic sectors. In the United States, agencies of the federal government and both houses of Congress typically respond by making massive amounts of relief available to the affected areas. This generally takes the form of short-term emergency measures to agricultural producers, such as feed assistance for livestock, drilling of new wells and low-interest farm operating loans. In the section below, the primary features of drought policy in the United States and Australia are compared. In addition, the approaches taken historically by Brazil and India are described briefly.

United States and Australia
Wilhite (1986) compared drought policy in the United States and Australia to learn more about the approaches taken by two drought-prone nations to deal with the effects of drought. For that study, the principal features of drought policy were grouped into three categories: organizational, response and evaluation (see Table 1).

Organizational features are planning activities that provide timely and reliable assessments, such as a drought early-warning system, and procedures for a coordinated and efficient response, such as drought declaration and revocation. These characteristics would be the foundation of a provincial, regional or national drought plan. Response features refer to assistance measures and associated administrative procedures that are in place to assist individual citizens or businesses experiencing economic and physical hardship because of drought.

Numerous assistance measures are available in the United States but few are intended specifically for drought. Table 2 lists the federal assistance programs used in the United States during the 1976–7 drought. Until recently, relief arrangements in Australia were included, for the most part, under the Natural Disaster Relief Arrangement agreements. These are now in the process of being discontinued as part of a proposed new national drought policy that will be discussed in a later section of this paper. Relief measures, by state, used during the 1982–3 severe drought in Australia are illustrated in Table 3.

Evaluation of organizational procedures and drought assistance measures in the post-drought recovery period is the third category of drought policy features. It is critical that governmental response efforts be evaluated during the post-drought period in order to avoid repeating the same mistakes during subsequent droughts. This evaluation is best accomplished by a nongovernmental organization, such as a university or private research group, that will be unbiased in its assessment. In Australia, governments have been more conscientious in their evaluation of drought response efforts. In the United States, the federal government has not routinely evaluated the performance of response-related procedures or drought assistance measures. Aspects of the 1976–7 drought were evaluated by the General Accounting Office (1979) and Wilhite et al. (1986). Responses to the 1987–9 droughts were examined by Riebsame et al. (1990).

Brazil
The most drought-prone region of Brazil is in the Northeast, often referred to as the 'drought polygon.' This region has a long history of drought, and the government has followed a variety of approaches to the problem, dating back to the Imperial Inquiry Commission that responded to the drought of 1877–9. One of the positive steps taken early to deal with the problem was the creation of the Department of Works to Overcome Drought (DNOCS) in 1909 (Pessoa 1987). Its purpose was to collect basic information about the region, including technical-scientific studies and maps, and to establish a meteorological and hydrological network for monitoring climate and water resources. In the 1960s, the Superintendency for Northeast Development (SUDENE) was created to expand existing monitoring networks, conduct hydrogeological research and integrated studies of potential natural resources, and map soil and mineral resources.

Table 1. *Comparison of drought policy features as of 1984: United States and Australia*

Features	United States	Australia
Organization		
National drought plan	None	Study in progess
State drought plans	In selected states	Through NDRA agreements
National drought early-warning system	Joint USDA/NOAA Weather Facility	Bureau of Meteorology
Agricultural impact assessment techniques	Available, but generally unreliable	Not available
Responsibility for drought declaration	Federal	State
Geographic unit of designation	County	Unit varies between states
Declaration procedures	Standard for all states; varies by program/agency	Varies between states; standard within states
Response		
State fiscal responsibility for assistance measures	Negligible, if any	Defined by NDRA agreements up to base amounts; varies by state
State administrative responsibility for assistance measures	No responsibility for federal measures	Defined by NDRA agreements and by federal measures
Eligibility requirements and provisions of drought assistance measures	Standard within programs for all designated counties	Varies by state for NDRA core measures; standard for federal programs
National crop insurance program	All-risk federal program	Rainfall insurance feasibility study in progress
Evaluation		
Post-drought documentation and evaluation of procedures and measures	No routine evaluation by government	Routine evaluation by federal and state governments

Source: NDRA, National Disaster Relief Arrangements; USDA/NOAA, US Department of Agriculture/National Oceanic and Atmospheric Administration.

In spite of the long history of actions taken to respond to drought in Northeast Brazil, the severe drought of 1979–83 found the region even more vulnerable to water shortages (Pessoa 1987). As a result, in 1985 the Civil Defense Plan was developed under the leadership of SUDENE to address both drought and flood problems. The purpose of the plan was to reduce the risks and impacts to the population and provide aid as necessary. The plan also triggers a drought watch system that produces more detailed climatological analyses and advisories.

Assistance programs have been of two types (Pessoa 1987). First, rural credit, water supply and food distribution programs are expanded to meet the needs of the distressed area. Second, public works projects are initiated to employ rural refugees in a variety of tasks, including:

- building water structures;
- transporting water supplies via tank trucks;
- providing reasonably priced staple food items;
- distributing food to ease social tension;
- planting trees;

- distributing fodder;
- supplying seeds;
- supporting small irrigation operations;
- distributing construction equipment;
- supporting literacy programs.

As a result of continuing problems in responding effectively to drought in the region, the government supported the conduct of a regional training seminar on drought management and preparedness in which the University of Nebraska's International Drought Information Center was one of the organizers and participants. As a result of this seminar, FUCEME (the State Meteorological Foundation of Ceará) is leading an effort aimed at enhancing regional coordination on drought planning.

India

Drought and famine mitigation efforts have had a long history in India, beginning with the adoption of 'famine codes' by several provincial governments in 1883 (Sinha *et al.* 1987). In 1975, the 'Drought Code' and 'Good Weather

Table 2. *Drought-related federal assistance programs used to respond to the 1976–7 drought in the United States, by agency*

Agency	Program name
Department of Agriculture	
Farmers Home Administration (FmHA]	Emergency Loans[a]
	Emergency Livestock Loans
	Farm Operating Loans
	Farm Ownership Loans
	Soil and Water Loans
	Irrigation and Drainage Loans
	Community Program Loans
Agricultural Stabilization and Conservation Service (ASCS)	Emergency Conservation Measures
	Emergency Livestock Feed
	Agricultural Conservation[a]
	Disaster Payments
Federal Crop Insurance Corp (FCIC)	Federal Crop Insurance[a]
Forest Service (FS)	Cooperative Forest Fire Control
	Cooperative Forest Insect and Disease Management
	Rural Community Fire Protection
	Drought-Related Stewardship
Soil Conservation Service (SCS)	Great Plains Conservation
	Resource Development and Conservation
	Conservation Technical Assistance
	Watershed Protection and Flood Prevention
Department of the Interior	
Bureau of Reclamation (BuRec)	Emergency Fund
	Drought Emergency[a]
	Drought-Related Technical Assistance
Bureau of Land Management (BLM)	Grazing Privilege
	Drought-Related Stewardship
Fish and Wildlife Service (FWS)	Drought-Related Stewardship
Southwest Power Administration	Emergency Electric Service[a]
Economic Development Administration (EDA), Department of Commerce	Community Emergency Drought Relief
	Economic Adjustment
	Public Works Impact Projects
Small Business Administration (SBA)	Emergency Drought Disaster Loans[a]
	Physical Disaster Loans
	Economic Injury Disaster Loans
Federal Disaster Assistance Administation (FDAA), Department of Housing and Urban Development	Disaster Assistance (Hay Transportation, Cattle Transportation, Emergency Livestock Feed, Forest Fire Suppression)
Federal Power Commission/Federal Energy Administration (FPC/FEA)	Drought-Related Services and Activities
Employment and Training Administation (ETA), Department of Labor	Unemployment Insurance Grants to States
	Farm Workers
	Comprehensive Employment and Training Programs (CETA)
	Employment Services
General Services Administration (GSA)	Donation of Federal Surplus Personal Property
	Sale of Federal Surplus Personal Property
Defense Civil Preparedness Agency (DCPA) Department of Defense	Civil Defense-Federal Surplus Personal Property Donations

Note: [a] Programs in the White House drought package.
Source: WESTPO (1977).

Table 3. *Drought relief measures available in Australia under the Natural Disaster Relief Arrangements, by state, as of March 1983*

Measure	New South Wales	Victoria	Queensland	South Australia	Western Australia	Tasmania	Northern Territory
Concessional loans							
Carry-on loans to primary producers	*	*	*	*	*	*	*
(Maximum amount ranges from $20 000 to $40 000, with interest at 4%. Repayment period generally 7 years with discretional repayment holiday of 1–3 years in some cases.)							
Restocking loans to primary producers	*	(1)	*	(1)	(2)	(1)	NA
(Maximum amount ranges from $20 000 to $30 000, repayable over 7–10 years, at 4–5% interest rate.)							
Loans for purchase of fodder	*	NA	NA	NA	NA	NA	NA
(Loans to dairy companies, repayable over 5 years, at 4% interest rate.)							
Loans for supply of water	NA	NA	(2)	NA	NA	NA	NA
(80% of cost to local authorities for augmentation of town water supplies. Repayable over 7–9 years at 3–4% interest rate.)							
Carry-on loans for small business	NA	*	(2)	*	*	NA	NA
(Maximum amount of $40 000, repayable over 7–10 years at 4% interest rate.)							
Loans to cereal growers	(2)	NA	NA	NA	(2)	NA	NA
Freight concessions							
Stock movement	*	*	*	*	*	NA	*
(Applies to rail and road at 75%.)							
Fodder	*	*	*	*	*	NA	*
(Applies to rail and road, generally at 50–75% concession.)							
Water to primary producers	*	*	*	*	NA	NA	NA
(Applies to private vehicle, generally at 75% concession.)							
Water to state, local or semigovernment authorities	NA	*	*	*	*	NA	NA
Machinery and equipment	NA	NA	(2)	NA	NA	NA	NA
Stock slaughter subsidy for primary producers	(2)	NA	(2)	(2)	(2)	(2)	(2)
(Generally $10–15 per head for cattle and $1–3 per head for sheep.)							
Stock disposal subsidy to local, state and semigovernment authorities	*	*	*	*	*	NA	NA
(Generally $1 per head for cattle and 15 cents per head for sheep.)							
Other subsidies							
Water	*	*	(2)	*	(2)	NA	NA
(Generally applies to drilling wells for towns or stock water at 75–100% concession.)							
Agistment	NA	(2)	(2)	NA	(2)	(2)	NA
(Rate of $1.00–1.75 per head of cattle and 10–12.5 cents per head for sheep and/or 50–75% of cost of adjustment.)							
Other	NA	(2)	(2)	NA	(2)	NA	NA

Notes:

*, included in core measures; NA, not available; (1), included in carry-on loans; (2), available but not part of core measures.

Code' were adopted. The Drought Code is anticipatory, providing a list of alternative cropping strategies that should be adopted when there is evidence of drought. These include anticipating conditions of food scarcity early in the season, maximizing production and alternating cropping patterns in irrigated areas, making mid-season corrections in crop planting in unirrigated areas, and building up seed and fertilizer buffers to implement the drought coping strategy. The Good Weather Code outlines the scientific, administrative and planning steps necessary to take full advantage of a good monsoon season to increase production of food grains. The Drought Watch group exists at the national level, made up of representatives of the Ministry of Agriculture, Meteorology Department, Indian Council of Agricultural Research, and Ministry of Information and Broadcasting, to monitor weather conditions throughout the country. This group receives regular reports from similar groups at the state and district levels.

The strategies currently being used by the Indian government to reduce vulnerability are a combination of emergency and long-term programs. These tactics include early monsoon forecasts; improved communication systems; provision of resources such as credit, fertilizers, pesticides, and power for increasing production; assistance to farmers in poor monsoon years; maintenance of adequate prices; maintenance of reasonable buffer stocks of food grains in strategic locations; and improved transportation systems (Sinha *et al.* 1987). The government has also undertaken a nationwide satellite monitoring program to provide early warning of the potential impacts of drought on agricultural production (Thiruvengadachari 1991). Evidence would seem to indicate that the drought-prone areas of India are less vulnerable to drought today than they were several decades ago because of the country's maintenance of buffer stocks of food for distribution during times of shortage (A. R. Subbiah, personal communication).

The Agro-Meteorology Service of India is striving to improve weather predictions, prepare climatological information for agricultural decision making, develop delivery systems to provide timely collection and distribution of data and information to users, and develop advisories on agricultural operations for contingency cropping practices during drought.

RECENT PROGRESS IN DROUGHT PREPAREDNESS

Governments worldwide have shown increased interest in drought planning since the early 1980s. Several factors have contributed to this interest. First, the widespread occurrence of severe drought over the past several decades and, specifically, the years during and following the extreme ENSO event of 1982–3 focused attention on the vulnerability of all nations to drought. Second, the costs associated with drought are now better understood by some governments. These costs include not only the direct impacts of drought but also the indirect costs (i.e. personal hardship, the costs of response programs, retardation of economic development and accelerated environmental degradation). Nations can no longer afford to allocate scarce financial resources to short-sighted response programs that do nothing to mitigate the effects of future droughts. Finally, the intensity and frequency of extreme meteorological events such as drought may increase, given projected changes in climate associated with increasing concentrations of carbon dioxide and other atmospheric trace gases. Droughts are a climatic certainty and recent events worldwide have highlighted the importance of preparing now for future episodes. From an institutional point of view, learning today to deal more effectively with climatic events such as drought may serve us well in preparing proper response strategies to long-term climate-related issues.

Global concern exists within the scientific and policy communities about the inability of governments to respond to drought in an effective and timely manner. In the past decade, numerous 'calls for action' for improved drought planning and management have been issued by national governments, professional organizations, international organizations, and others. The challenge of changing the perception of policy makers and scientists worldwide about drought is a formidable one. The typical mode of operation for government in dealing with natural hazards is crisis management. It is indeed a difficult task for government to engage in long-range planning. However, the progress currently being made in planning for future drought demonstrates a new awareness and improved understanding of drought and its economic, social and environmental impacts.

United States

In the past decade, droughts have been a prevalent feature of the American landscape. These droughts have resulted in significant impacts in a myriad of sectors, including agriculture, transportation, energy, recreation and health; they have also had adverse environmental consequences. In society's attempt to cope with the effects of these extended periods of water shortage in recent years, the inadequacy of federal contingency planning efforts has been confirmed once again. The inability of the United States government to respond effectively has also illustrated the inflexibility of existing water management systems and policies as well as the lack of coordination between and within levels of government.

Previous studies have demonstrated that the impacts of both short-term and multi-year drought in the United States have been aggravated by poorly conceived or nonexistent assessment and response efforts by governments. These efforts have been characterized as largely ineffective, poorly coordinated, and untimely (General Accounting Office 1979; Wilhite *et al.* 1986; Wilhite and Easterling 1987). As a result, there have been numerous 'calls for action' by regional and national organizations for the development of a national drought policy to coordinate federal response to drought. These calls include recommendations from the Western Governors' Policy Office (1978), General Accounting Office (1979), National Academy of Sciences (1986), Interstate Conference on Water Policy (1987), Environmental Protection Agency (Smith and Tirpak 1989), Great Lakes Commission (1990), and the American Meteorological Society (Orville 1990). The call from the Environmental Protection Agency (EPA) has come about as a result of the concern that exists about a possible increase in the frequency and severity of extreme events in association with projected changes in climate because of increasing concentrations of atmospheric trace gases.

Despite the numerous calls for the development of a national drought policy and plan, the federal government has not acted on these recommendations. The primary reason for the lack of progress by federal agencies seems to be the unique character and multidisciplinary nature of drought and the cross-cutting responsibilities of federal agencies for drought assessment and response programs. Clearly, a single federal agency must take the lead in coordinating the development of a national plan. It is less clear which federal agency should assume this responsibility. In the final analysis, it may take an executive order to initiate the process at this level. In the meantime, progress in drought management at the federal level has been sluggish and agency-specific (e.g. Corps of Engineers, Bureau of Reclamation).

Because of the factors mentioned above and an apparent lack of appreciation by federal agencies of the complexity and seriousness of drought management issues faced by states, it became clear to many states during the mid-1980s that progress toward a higher level of preparedness would be achieved only if they took the lead. Historically, state governments have played a passive role in governmental efforts to assess and respond to drought. During the widespread and severe drought of 1976–7, for example, no state had prepared a formal drought response strategy. In 1982, only three states had developed plans: South Dakota, New York and Colorado. Generally speaking, states have relied on the federal government to come to their rescue when water shortages reach near-disaster proportions by providing relief to drought victims.

At present, 23 states have prepared some type of formal drought contingency plan (Wilhite 1991*a*). The pattern of states with drought plans is illustrated in Fig. 1. This pattern is complex and can be only partially explained on the basis of the climatology of drought. Impediments to plan develop-

ment were discussed earlier in the chapter. However, each state's decision to develop (or not to develop) a drought plan is based on specific climatological, political, economic and demographic factors. An analysis of the relative importance of these factors has been completed but will not be discussed here (Wilhite and Rhodes 1994). For those states that have developed plans, planning efforts have often been conducted in conjunction with an overall water management planning initiative. Clearly, states can now be labelled policy innovators in drought planning.

An examination of existing state drought plans reveals that they have certain key elements in common. Administratively, a task force is responsible for the operation of the system and is directly accountable to the governor. The task force keeps the governor advised of water availability and potential problem areas; it also recommends policy options for consideration. Operationally, drought plans have three features in common. First, a water availability committee continuously monitors water conditions and prepares outlooks a month or season in advance. Since most of the information necessary to monitor water conditions comprehensively (i.e. precipitation and temperature, streamflow, groundwater levels, snowpack, soil moisture, meteorological forecasts) is available from state or federal agencies, the primary role of the committee is to coordinate the collection and analysis of this information and the delivery of products to decision makers on a timely basis. The committee assimilates this information and issues timely reports and recommendations. Second, a formal mechanism usually exists to assess the potential impacts of water shortages on the most important economic sectors. In some states this task is accomplished by a single committee or, more commonly, separate working groups are established to address each sector. Third, a committee or the task force referred to previously considers current and potential impacts and recommends response options to the governor.

Although many of the mitigative programs implemented by states during recent droughts can be characterized as emergency actions taken to alleviate the crisis at hand, these actions were often quite successful. As states gain more experience assessing and responding to drought, future actions will undoubtedly become more timely and effective. State drought preparedness plans will become broader in scope, addressing a wider range of potential mitigative actions, including more meaningful levels of intergovernmental coordination. In time this will help states avoid or reduce the impacts, conflicts and personal hardships associated with drought. To be successful, these plans should be integrated with local, regional and national plans, if they exist.

Fortunately, many resources are now available to assist governments in the drought planning process. The existence of model plans (Western States Water Council 1987; Wilhite

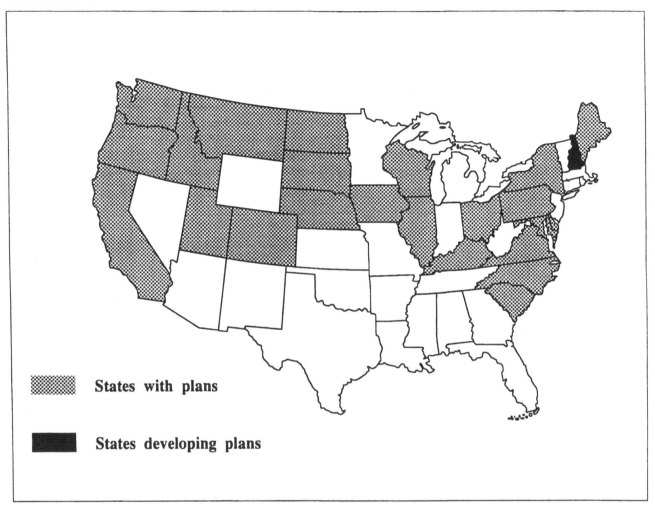

States with plans

States developing plans

Figure 1: States in the contiguous United States with drought plans, as of 1991. (Alaska and Hawaii have not prepared drought plans.)

1990) and 23 state plans provide a critical reference for states desiring to develop a plan or revise an existing plan. In addition, several regional organizations have considerable experience in drought planning and can assist states in plan development (e.g. Delaware River Basin Commission, Great Lakes Commission, Western Governors' Association).

South Africa

Actions taken by the South African government in response to droughts have typically been poorly coordinated and assistance programs have been largely ineffective (C. R. Baard, personal communication 1985). According to Baard, the government has had difficulty assessing drought impact and making subsequent declarations, and no routine comprehensive evaluation of government drought policy and response efforts has been completed.

For many decades, drought assistance programs in South Africa concentrated mainly on providing relief to the livestock industry, with little attention to crop farming, either

dryland or irrigated (Wilhite 1987). The rationale behind this emphasis on the livestock industry in South Africa has been that 85% of all agricultural land in the country remains under native pastures, most of which lie in the dry zones of the western and northwestern part of the country. The incidence of drought in these drier zones is about one year in three. Only 15% of South Africa receives precipitation in excess of 500 millimeters per year. A serious drought that began in 1978 and affected, to varying degrees, 75% of South Africa resulted in significant expenditures by the government for drought relief. For example, during the 1984–5 fiscal year the government spent approximately R447 million in support of various relief programs (C. R. Baard, personal communication 1985). During the years 1987–9 the government allocated R1300 million to drought and flood relief schemes (Bruwer 1990). Expenditures of this magnitude represent a significant expenditure of funds and illustrate the serious threats that natural disasters pose to the country.

In the decades immediately preceding the 1980s, drought relief was provided through a phased approach, but only to farmers in those areas officially designated by the government (Wilhite 1987). The principal purpose of these assistance programs was to help livestock farmers preserve their herds until dry conditions eased. This assistance was intended to apply only to extended or disaster droughts, although it was often difficult to distinguish between these and 'normal' droughts. Assistance provided was generally in the form of rebates (phase 1) for transportation costs incurred in importing livestock feed to the affected area or in shipping animals to areas where grass was available. If drought conditions continued to deteriorate, loans to purchase livestock feed (phase 2) were then made through the Agricultural Credit Board. A continuation of drought conditions brought about the availability of subsidies from the government to farmers to help pay for feed (phase 3). One of the principal difficulties with this phased approach was that it did not encourage farmers to adopt production strategies that favored a minimization of risk to the agricultural resource base (soil, water and vegetation), an approach more in harmony with environmental constraints (Bruwer 1990). Indeed, farmers prefer to strive for maximum production, regardless of the potential effects on the resource base.

After 1980, the drought relief scheme was modified, placing greater emphasis on the preservation of the agricultural resource base and the self-sufficiency of livestock farmers to endure droughts of other than disaster proportions (Bruwer 1990). The current approach requires a reduction in stock numbers as a prerequisite for eligibility for the forms of relief available during a 'disaster' drought. In order to facilitate this approach, the country was divided into grazing capacity zones. Grazing capacity is defined as the number of hectares per livestock unit which can be kept and maintained on the natural veld or grassland, as well as planted pastures, crop residues and any other fodder produced on the farm. This new relief scheme provided for rebates on the transportation costs of livestock feed, incentives for stock reduction, loans and subsidies for the cost of livestock feeds in order to maintain the herd nucleus, and subsidies for finishing stock in feedlots. Incentives were in the form of monthly payments to farmers and were calculated on a per livestock unit basis. Consideration was given to the type of stock (i.e. large versus small) in the calculation of incentives. Other types of assistance now available to farmers during droughts include a water quota subsidy for irrigators and incentives for converting marginal cultivated lands to perennial pasture crops in both summer and winter rainfall zones.

To administer the new drought policy and relief scheme an institutional structure was established. This structure included a National Drought Committee (NDC), with multiagency representation, to advise the Minister of Agriculture on drought assistance matters and to scrutinize applications for assistance from affected areas (Bruwer

1989). District Drought Committees (DDC) were also established at the local level to consider all applications for the designation or revocation of disaster drought areas according to the criteria specified by the NDC. The NDC is responsible for approving or rejecting these applications. The DDC is composed of the magistrate (chair) and representatives of the District Farmers' Union, Agricultural Credit Committee, Soil Conservation Committee, and the Department of Agriculture and Water Supply.

On the basis of experiences with the new drought policy during the 1980s, the government of South Africa is convinced that the new relief scheme has contributed significantly to sustained agricultural production and development, helped to maintain rural communities and infrastructure, counteracted unemployment, reduced political pressure, and increased cooperation between agricultural groups and government, thus promoting mutual acceptance of responsibility for coping with disasters (Bruwer 1990). However, Bruwer has noted some deficiencies and shortcomings of the current system. These include the lack of adequate indices to identify disaster droughts, lack of suitable assessment procedures and inadequate monitoring techniques (including an improved weather station network). A considerable amount of drought-related research also needs to be undertaken, including post-drought audits of past relief efforts.

To assist the DDCs with the evaluation of drought intensity and the determination of eligibility for drought relief, the government recently implemented a scheme that provides for greater uniformity, objectivity and accuracy in the assessment of drought impact. The main elements evaluated by the procedure are climate, veld, pastures and crops, livestock and water (Roux 1991).

The process of developing a better approach to drought management in South Africa is not complete. The government continues to strive for better ways to reduce the risk of drought through proactive measures. According to Bruwer, 'society is demanding a more rational, cost effective and proactive approach' for future drought relief schemes. It is essential that this approach reduce the taxpayer's burden and provide incentives for diminishing natural resource degradation.

Australia

The Australian constitution does not delegate specific powers covering natural disaster relief to the federal government. These powers belong primarily to the states, which, as a result, have taken a more active role in drought response than state governments in the United States and elsewhere.

Before 1971, natural disaster relief and restoration was provided at a state's request by joint federal/state financing

Table 4. *Expenditures in Australian states under Natural Disaster Relief Arrangements, by type of disaster, 1970–1 to 1983–4 (A$ thousands)*

	New South Wales	Victoria	Queensland	South Australia	Western Australia	Tasmania	Northern Territory	Total
1970–1	3239	—	15623	—	—	596	—	19458
1971–2	458	—	3143	—	—	—	—	3601
1972–3	—	—	—	—	—	—	—	—
1973–4	987	—	—	—	—	—	—	987
1974–5	160	—	—	—	—	—	—	160
1975–6	—	—	—	—	—	—	—	—
1976–7	1120	1626	—	—	3023	—	—	5769
1977–8	2620	1228	2785	13580	17999	—	—	38212
1978–9	3013	1422	5165	9257	8070	—	—	26927
1979–80	—	—	2208	2225	12560	—	—	16993
1980–1	66810	—	22768	—	20142	—	—	109720
1981–2	31018	—	9608	—	5081	295	—	46002
1982–3	53645	34796	51982	27380	12653	1282	—	181738
1983–4 (estimate)	21500	8100	63300	4600	22100	1900	—	121500
Total	184570	47172	176582	57042	101628	4073	—	571067

Source: National Drought Consultative Committee (1984).

on a one-to-one cost-sharing basis. No limit was set on the level of funding that could be provided by the federal government. In 1971 the Natural Disaster Relief Arrangements (NDRA) were established, whereby states were expected to meet a certain base level or threshold of expenditures for disaster relief from their own resources (Department of Primary Industry 1984). Disasters provided for in this arrangement are droughts, cyclones, storms, floods and bush fires. These expenditure thresholds were set according to 1969–70 state budget receipts and therefore varied between states. The base levels were raised in 1978 and 1984 (National Drought Consultative Committee 1984; Keating 1984). Under the NDRA system, the federal government agreed to provide full reimbursement of eligible expenditures after the thresholds for state expenditures on natural disasters were reached. The NDRA formalized, for the first time, joint federal–state natural disaster relief arrangements.

At the time of the establishment of NDRA, a special set of core measures (i.e. federal government-approved drought assistance measures) had evolved in each state on the basis of 30 years of government involvement in disaster relief. These measures were particularly relevant to the needs of each state because they had been designed by state government in response to its own disaster-related experiences.

Tables 4 and 5 provide data on state and federal expenditures for drought aid from 1970–1 to 1983–4 under the NDRA. The magnitude of state expenditures is significant,

especially when compared with the limited financial responsibility of states in the United States. The total for all states was just over A$570 million. Of this total, approximately A$180 million was expended during 1982–3 and A$120 million during 1983–4. Federal expenditures to the states for drought aid under the NDRA arrangements (Table 5) were just under A$370 million, or about A$200 million less than the total state expenditures. The largest share of the assistance was provided to Queensland and New South Wales.

In addition to the cost-sharing measures described above, two federal drought assistance schemes were available during the 1982–3 drought. These were the Drought Relief Fodder Subsidy Scheme and the Drought Relief Interest Subsidy Scheme (National Drought Consultative Committee 1984). The Fodder Subsidy Scheme provided a payment to drought-declared primary producers to help defray the cost of fodder for sheep and cattle. The administrative costs of this program were covered by the states. The amount of the subsidy was based on 50% of the price of feed wheat and the nutritive value of the fodder relative to wheat; Commonwealth expenditures under this program were about A$104 million during 1982–3 and A$18 million through February 1984.

The Drought Relief Interest Subsidy Scheme provided payments to eligible primary producers to cover all interest payments exceeding 12% per year. To be eligible, producers had to have been drought declared and could not have

Table 5. *Commonwealth of Australia payments under Natural Disaster Relief Arrangements, estimated by type of disaster, 1970–1 to 1983–4 (A$ thousands)*

	New South Wales	Victoria	Queensland	South Australia	Western Australia	Tasmania	Northern Territory	Total
1970–1	450	—	13 632	—	—	16	—	14 098
1971–2	—	—	1 502	—	—	—	—	1 502
1972–3	—	—	46	—	—	—	—	46
1973–4	38	—	—	—	—	—	—	38
1974–5	114	—	—	—	—	—	—	114
1975–6	—	—	—	—	—	—	—	—
1976–7	779	716	—	—	2 134	—	—	3 629
1977–8	1 458	399	3 091	12 350	15 269	—	—	32 567
1978–9	743	173	2 942	5 430	6 036	—	—	15 324
1979–80	—	− 229	1 224	− 270	6 922	—	—	7 647
1980–1	42 447	—	14 780	− 737	13 523	—	—	70 013
1981–2	14 554	—	5 162	—	2 239	267	—	22 222
1982–3	32 557	22 695	37 297	18 368	7 731	—	—	118 648
1983–4 (estimate)	11 800	4 600	45 300	4 300	15 300	600	—	81 900
Total	104 940	28 354	124 976	39 441	69 154	833	—	367 748

Source: National Drought Consultative Committee (1984).

available financial assets in excess of 12% of the total farm debt. Expenditures for the program, not including administrative costs, were about A$3 million in 1982–3 and A$23 million through February 1984.

The Livestock and Grain Producers Association (LGPA) of New South Wales strongly commended the state and federal governments of Australia for their drought assistance measures. LGPA based its conclusions on the achievement of what it considers to be the first priority of drought aid in Australia – the preservation of the national sheep and cattle herd. Through the preservation of these resources, farm and non-farm income was able to recover more quickly than after previous episodes of severe drought. LGPA estimated that, had government not intervened in 1982–3, some 15–20 million sheep would have been slaughtered. As a result, post-drought recovery would have been delayed, at a cost to the national economy of A$500 million over a 5-year period (Anonymous 1983). However, the Australian Agricultural Council (1983) concluded, 'With the exception of concessional finance and information, existing policy measures, including those introduced during the current (1982–83) drought, do not perform well in achieving the objectives of drought policy which it considered important. In summary, the nearly $300 million of expenditures was not cost effective.'

These contrasting views of the cost effectiveness of recent drought measures in Australia reflect the recent controversy over state and federal involvement in drought aid. Several other studies have been completed (National Farmers' Federation 1982; South Australian Department of Agriculture 1983; Stott 1983), each providing recommendations for future drought policy. A National Drought Consultative Committee (NDCC) was appointed by the Minister for Primary Industry in 1984 to review Australian drought policy.

In April 1989 the Commonwealth government decided to remove drought from the NDRA scheme described previously. Following this action, a drought policy review was recommended by the Commonwealth in May 1989 under the leadership of the Minister for Primary Industries and Energy. The objectives of this review (Drought Policy Review Task Force 1990) were to (1) identify policy options that encourage primary producers and other segments of rural Australia to adopt self-reliant approaches to the management of drought; (2) consider the integration of drought policy with other relevant policy issues; and (3) advise on priorities for Commonwealth government action in minimizing the effects of drought in the rural sector. An important aspect of this policy review was to examine the extent to which the policies of the Commonwealth government promote more effective farm management given the seasonality of climates and climatic variability. The task force concluded that the relief measures that have been used in the past have not provided a positive incentive for effective farm management or responsible land management. On the contrary, it was determined that common misperceptions of drought have guided past policies by government, leading to a process of crisis management or 'gambling on the weather' by the agricultural community (Drought Policy Review Task Force 1990).

Several objectives of a newly defined national drought policy emerged from the task force review. These objectives are to (1) encourage primary producers and other segments of rural Australia to adopt self-reliant approaches in managing for climatic variability; (2) facilitate the maintenance and protection of Australia's agricultural and environmental resource base during periods of increasing climate stress; and (3) facilitate the early recovery of agricultural and rural industries consistent with long-term sustainable levels. Within this framework, numerous more specific objectives of these policies were stated. The primary thrust of this change in national policy is from one of crisis management to one of risk management. The intent of the task force was to apply this approach at two management levels: farm and government policy. This integrated approach is the foundation of the proposed changes in national policy, changes that have met with some resistance. The proposed changes are presently under review.

ADVANCING DROUGHT PREPAREDNESS IN THE 1990s

Drought policy versus planning objectives

Drought planning is defined as actions taken by individual citizens, industry, government, and others in advance of drought for the purpose of mitigating some of the impacts and conflicts associated with its occurrence (Wilhite 1991b). From an institutional perspective, a drought preparedness plan should include, but is not limited to, the following elements:

(1) A comprehensive, integrated monitoring/early warning system to provide decision makers at all levels with information about the onset, continuation and termination of drought conditions and their severity.
(2) Operational impact assessment programs to determine reliably the likely effects of drought in a timely manner.
(3) An institutional structure for coordinating governmental actions, including information flow within and between levels of government, and drought declaration and revocation criteria and procedures.
(4) Appropriate drought assistance programs (both technical and relief) with predetermined eligibility and implementation criteria.
(5) Financial resources to maintain operational programs and to initiate research required to support drought assessment and response activities.

(6) Educational and public awareness programs designed to promote an understanding and adoption of appropriate drought mitigation and water conservation strategies among the various economic sectors most affected by drought.

To be successful, drought planning must be integrated between levels of government, involving the private sector, where appropriate, early in the planning process. As we have seen from the discussion presented in the previous section, progress has been made by some governments in taking a more proactive approach to drought management. For the majority of nations, however, much needs to be done. This final section of this chapter presents some key factors that should be considered by governments attempting to adopt the risk management approach in future drought response efforts.

Prior to the development of a drought preparedness plan, government officials should formulate a drought policy to define what they hope to achieve with that plan (Wilhite 1991b). The objectives of a drought policy differ from those of a drought plan. There must be a clear distinction of these differences at the outset of the planning process. A drought policy will be broadly stated and should express the purpose of government involvement in drought assessment, mitigation and assistance programs. Drought plan objectives are more specific and action oriented. Typically, the objectives of drought policy have not been stated explicitly by government. What generally exists in many countries, including the United States, is a *de facto* policy, one defined by the most pressing needs of the moment. Ironically, under these circumstances it is the specific instruments of that policy (such as assistance measures, including grants and low-interest loans, and so forth), particularly at the federal level, that define the objectives of the policy. Without clearly stated drought policy objectives, the effectiveness of assessment and response activities is difficult to evaluate.

The objectives of drought policy should encourage or provide incentives for agricultural producers, municipalities, and other water-dependent sectors or groups to adopt appropriate and efficient management practices that help to alleviate the effects of drought. Past relief measures have, at times, discouraged the adoption of appropriate management techniques. Assistance should also be provided in an equitable, consistent and predictable manner to all without regard to economic circumstances, industry or geographic region. Assistance can be provided in the form of technical aid or relief measures. Whatever the form, those at risk would know what to expect from government during drought and thus would be better prepared to manage risks. At least one objective should also seek to protect the natural and agricultural resource base. Degradation of these resources can result in spiralling economic, environmental and social costs.

The objectives of drought policy can be achieved only if they are formulated at the initiation of the planning process. The entire planning process can then be structured around these basic themes. One question that government officials must address is the purpose and role of government involvement in drought mitigation efforts. Other questions should address the scope of the plan; identification of geographic areas, economic sectors and population groups that are most at risk; principal environmental concerns; and potential human and financial resources to invest in the planning process. Answers to these and other questions should help to determine the objectives of drought policy and therefore provide a focus for the drought planning process.

Constraints to drought planning

Institutional, political, budgetary and human resources constraints often make drought planning difficult (Wilhite and Easterling 1987, 1991). One major constraint that exists worldwide is a lack of understanding of drought by politicians, policy makers, technical staff and the general public. Lack of communication and cooperation among scientists, and inadequate communication between scientists and policy makers, on the significance of drought planning, also complicate efforts to initiate drought planning. Because drought occurs infrequently in some regions, governments may ignore the problem or give it low priority. Inadequate financial resources to provide assistance and competing institutional jurisdictions between and within levels of government may also serve to discourage governments from undertaking drought planning. Other constraints include technological limits such as difficulties in predicting and detecting drought, insufficient data bases, and inappropriate mitigation technologies.

Policy makers and bureaucrats should understand that droughts, like floods, are a normal feature of climate. Their recurrence is inevitable. Drought manifests itself in ways that span the jurisdiction of numerous bureaucratic organizations (e.g. agricultural, water resources, health) and levels of government (e.g. federal, state and local). Competing interests, institutional rivalry and 'turf protection' impede the development of concise drought assessment and response initiatives. To solve these problems, policy makers and bureaucrats, as well as the general public, must be educated about the consequences of drought and the advantages of preparedness. Drought planning requires input by several disciplines, and decision makers must play an integral role in this process.

The development of a drought plan is a positive step that demonstrates governmental concern about the effects of a potentially hazardous and recurring phenomenon. Planning, if undertaken properly and implemented during non-drought periods, can improve governmental ability to respond in a timely and effective manner during periods of crisis. Thus, planning can mitigate and, in some cases, prevent some impacts while reducing physical and emotional hardship. Planning is a dynamic process that must incorporate new technologies and take into consideration socioeconomic, agricultural and political trends.

It is sometimes difficult to determine the benefits of drought planning versus the costs of drought. There is little doubt that drought preparedness requires financial and human resources that are, at times, scarce. This cost has been and will continue to be an impediment to the development of drought plans. Preparedness costs are fixed and occur now while drought costs are uncertain and will occur later. Further complicating this issue is the fact that the costs of drought are not solely economic. They must also be stated in terms of human suffering and the degradation of the physical environment, items whose values are inherently difficult to estimate.

Post-drought evaluations have shown assessment and response efforts of state and federal governments with a low level of preparedness to be largely ineffective, poorly coordinated, untimely and economically inefficient. Unanticipated expenditures for drought relief programs are devastating to government budgets. For example, during the droughts of the mid-1970s in the United States, specifically 1974, 1976 and 1977, the federal government spent more than $7 billion on drought relief programs. As a result of the drought of 1988, the federal government spent $3.9 billion on drought relief programs and $2.5 billion on farm credit programs. A disaster relief package was also passed by the US Congress in August 1989 in response to a continuation of drought conditions. Between 1970 and 1984, state and federal government in Australia expended more than A$925 million on drought relief under the Natural Disaster Relief Arrangements. The Republic of South Africa has spent R2.5 billion on drought relief in the past decade. When compared with these expenditures, a small investment in mitigation programs in advance of drought would seem to be a sound economic decision. In developing countries, droughts devastate regional and national economies and significantly hinder the development process.

It is important to remind decision makers and policy officials that, in most instances, drought planning efforts will use existing political and institutional structures at appropriate levels of government, thus minimizing start-up and maintenance costs. It is also quite likely that some savings may be realized as a result of improved coordination and the elimination of some duplication of effort. Also, drought plans should be incorporated into general natural disaster and/or water management plans wherever possible. This would reduce the cost of drought preparedness substantially. Politicians and many other decision makers must simply be better informed about drought, its impacts, and alternative management approaches and how existing information and technology can be used more effectively to reduce the impact of drought at a relatively modest cost.

The development of a drought policy and plan: the ten-step process

A planning process was developed recently in the United States in order to facilitate the preparation of drought contingency plans by state government decision makers (Wilhite 1990, 1991). The proposed process is intended to assist government decision makers in improving drought mitigation efforts through more timely, effective and efficient assessment and response activities. The framework below presents the principal steps in the planning process in order for government to address its drought-related concerns. However, the process is intended to be flexible (i.e. governments can add, delete or modify steps as necessary).

The intent here is not to present a detailed discussion of each of these steps. What is included is a very brief description of the purpose and elements of each step as these relate to the overall planning process. This process must be modified or adapted to each region, adding or deleting steps as appropriate.

Step 1. Appointment of national/state drought task force or committee

The drought task force (DTF) or committee should be appointed by the president, governor or designated government official and include representatives from all relevant agencies of government. This task force will be composed of senior policy makers.

The DTF has two purposes. First, during plan development, the DTF will supervise and coordinate the development of the plan. Second, after the plan is implemented and during times of drought when the plan is activated, the DTF will assume the role of policy coordinator – reviewing and recommending alternative policy response options to the appropriate policy official. The makeup of the DTF should recognize the multidisciplinary nature of drought and its impacts and include representatives of both state and federal government. The DTF should consider including a representative of the media or a public information specialist in an advisory capacity so that the proper mechanisms are incorporated into the plan to ensure public awareness of drought severity and the actions implemented by government.

Environmental and public interest groups may also be included on the DTF, or they may serve in an advisory capacity. The actual makeup of the task force is expected to be highly variable between states or countries, reflecting the variety of economic sectors affected and political infrastructure. Membership should be kept relatively small so that size does not become an impediment to the planning process. What is envisioned is the development of an infrastructure that can not only assess and respond to short-term reductions in water supply due to drought, but also can address questions of changes in vulnerability in the long term.

Step 2. Statement of drought policy and planning objectives

The first official action of the DTF will be the determination of a drought policy. This policy will lead to the development of a general statement of purpose for the drought plan.

A general statement of purpose for a drought plan could be to provide an effective and systematic means of assessing and responding to drought conditions. The DTF then must identify specific objectives of the plan. Drought plan objectives and their applications will vary between countries or states, reflecting the unique physical, environmental, socioeconomic and political characteristics of each location. Some objectives that might be considered include:

(1) To provide timely and systematic data collection, analysis and dissemination of drought-related information.
(2) To establish proper criteria to identify and designate drought-affected areas of the state and to trigger the initiation and termination of various assessment and response activities by governmental agencies during drought emergencies.
(3) To provide an organizational structure that assures information flow between and within levels of government and defines the duties and responsibilities of all agencies with respect to drought. To ensure adequate coordination between the federal and state governments, this structure should be integrated with national drought policies (if they exist).
(4) To maintain a current inventory of assistance programs used in assessing and responding to drought emergencies and provide a set of appropriate action recommendations.
(5) To provide a mechanism to assure the timely and accurate assessment of drought impact on agriculture, industry, municipalities, wildlife, health, and other areas as appropriate.
(6) To provide accurate and timely information to the media in order to keep the public informed of current conditions and response actions.
(7) To establish and pursue a strategy to remove obstacles to the equitable allocation of water during shortages and to provide incentives to encourage water conservation.
(8) To establish a set of procedures to evaluate and revise the plan on a continuous basis in order to keep the plan responsive to the needs of the state.

*Step 3. Resolving conflict between environmental and
economic sectors*

Political, social and economic values often clash during
drought conditions as competition for scarce water resources
intensifies, and it may be difficult to achieve compromises. To
reduce the risk of conflict between water users during periods
of shortage, it is essential for the public to receive a balanced
interpretation of changing conditions through the media.
The DTF should ensure that frequent, thorough and accurate news releases are issued to explain changing conditions
and complex problem areas. To lessen conflict and develop
satisfactory solutions, it is essential that the views of citizens
and public and environmental interest groups be considered
in the drought planning process at an early stage. Although
the level of involvement of these groups will no doubt vary
notably, the power of public interest groups in policy making
is considerable. Public interest organizations have initiated
and participated in the development of natural resource
policies and plans for some time and have considerable
experience with this process. The involvement of these
groups in determining appropriate policy goals strengthens
the overall policy and plan. Moreover, this involvement
assures that the diverse values of society are adequately
represented in the policy and plan.

*Step 4. Inventory of natural, biological, and human resources
and financial and legal constraints*

The DTF should undertake an inventory of natural, biological and human resources, including the identification of
financial and legal constraints. Resources include, for example, natural and biological resources, human expertise,
infrastructure, and capital available to government. Financial constraints include costs of hauling water or hay, new
program or data collection costs, and so forth; legal constraints include user water rights, existing public trust laws,
methods available to control usage, requirements for contingency plans for water suppliers, and emergency and other
powers of the government during water shortages. An inventory of these resources would reveal assets and liabilities that
might have an effect on the planning process; in addition, a
comprehensive assessment of available resources would
provide the information necessary for further action by the
task force.

Step 5. Development of the drought plan

The DTF will be the coordinating body for the development
of a drought plan. The plan is envisioned to follow a stepwise
or phased approach as water conditions deteriorate and
more stringent actions are needed. Thresholds must be
established such that, when exceeded, certain actions are
triggered within government agencies, as defined by the

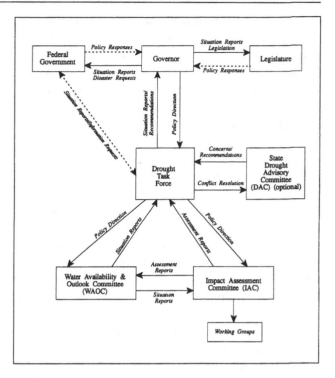

Figure 2: Linkages and suggested organizational components of
the drought plan.

structure of the plan. A flow chart illustrating these linkages
and the suggested components of the drought plan is shown
in Fig. 2.

A drought plan possesses three essential elements: monitoring, impact assessment and response. These elements are
the basis for three committees: (1) Water Availability and
Outlook Committee (WAOC); (2) Impact Assessment Committee (IAC); and (3) Drought Response Committee or
Drought Task Force (DTF). Although each committee has
its own distinct activities, formal linkages will need to be
incorporated in the plan for the committees to function
properly and be responsive to state needs and changing
conditions. The WAOC's activities would include defining
drought and developing triggers, identifying drought management areas, developing a monitoring system for drought,
completing an inventory of observation networks, determining primary users and their needs, and developing data and
information delivery systems. Membership of the committee
should include representatives from agencies with responsibilities for forecasting and monitoring the principal indicators of the water balance.

During periods of drought, impacts will be far-reaching
and cut across economic sectors and the responsibilities of
government agencies. The IAC will represent those economic
sectors most likely to be affected by drought. The IAC
chairperson should be a permanent member of the DTF; the
rest of the committee should consist of an interagency team

of agency heads or their representatives. The IAC should consider both direct and indirect losses resulting from drought, since its effects ripple through the economy. Because of the obvious dependency of the IAC on the WAOC, frequent communication is essential. What is recommended is a series of working groups responsible for anticipating and identifying drought-related impacts in each economic sector. The responsibility of the IAC is to coordinate the activities of each of the working groups and make policy response recommendations to the DTF.

A drought response committee comprising senior-level officials will act on the information and recommendations of the IAC and evaluate the state and federal programs available to assist agricultural producers, municipalities, and others during times of emergency. The makeup of this committee is envisioned to be roughly the same as that of the DTF. Therefore, for maximum efficiency the DTF can assume this function once the plan has been developed and fully implemented. The DTF will present its recommendations to the governor.

During the plan development process, the DTF should make an inventory of all forms of assistance available from government during severe drought and evaluate these programs for their ability to address short-term emergency situations and as long-term mitigation programs to reduce vulnerability to drought. The DTF should also be aware of the proper protocol for requesting federal assistance.

Step 6. Identification of research needs and institutional gaps
The purpose of this step is to identify research needed in support of the objectives of the drought plan and to recommend research necessary to remove deficiencies that may exist. The research needs and institutional gaps will be identified by the monitoring, impact assessment and response committees. These committees will make recommendations to the national/state drought committee for further action.

Step 7. Synthesis of scientific and policy issues
Direct and extensive contact is required between scientists and policy makers to distinguish what is feasible from what is desirable. Typically little contact occurs between these two groups. The purpose of this step is to identify ways to break down the barriers that exist between disciplines and between scientists and policy makers.

Step 8. Implementation of the drought plan
The drought plan should be implemented by the DTF to give maximum visibility to the program and credit to the agencies and organizations that have a leadership or supporting role in its operation. The plan should be tested under simulated drought conditions before it is implemented, and simulation exercises should be carried out periodically following implementation. It is also suggested that announcement and implementation occur just before the most drought-sensitive season to take advantage of inherent public interest.

Step 9. Development of multilevel educational and training programs
Educational and training programs must be long term in design, concentrating on a broad audience ranging from policy makers to extension personnel to individual citizens.

Educational and training programs should emphasize several points. First, a greater level of understanding must be established to heighten public awareness of drought and water conservation and the ways in which individual citizens, industry and government can help to mitigate impacts in the short run. This educational process might begin with the development of a media awareness program. Second, the DTF should initiate an information program aimed at educating the general population about drought and drought management and what they can do as individuals to conserve water in the short run. Educational programs must be long term in design, concentrating on achieving a better understanding of water conservation issues among elementary school children. If such programs are not developed, governmental and public interest in and support for drought planning will wane during periods of non-drought conditions.

Step 10. Development of drought plan evaluation procedures
The drought plan must be evaluated and revised periodically to remain responsive to the needs of each country. Two modes of evaluation are recommended. The first is a continuous (every 1–2 years) evaluation and revision to adjust the plan in light of political, economic, technological and social changes. This mode of evaluation is intended to express drought planning as a dynamic process, rather than a discrete event. The evaluation process is proposed to keep the drought assessment and response system current and responsive to the needs of society. Following the initial establishment of the plan, it should be monitored routinely to ensure that societal changes that may affect water supply and/or demand or regulatory practices are considered for incorporation.

The second mode of evaluation follows an episode of severe drought in which the plan was activated. A post-drought evaluation of the plan should be undertaken by a non-governmental organization to assure an unbiased appraisal of the assessment and response actions. Institutional memory fades quickly following drought as a result of changes in political administration, natural attrition of persons in primary leadership positions, and the destruction of critical documentation of events and actions taken.

CONCLUSION

Post-drought audits of government response to drought have demonstrated that the reactive or crisis management approach has led to ineffective, poorly coordinated and untimely responses. Also, the magnitude of economic, social and environmental losses in recent years has illustrated the vulnerability of all nations, developed and developing, to extended episodes of severe drought. Increased awareness and understanding of drought have led many governments to take a more proactive approach toward drought management, thus attempting to reduce impacts in the short term and vulnerability in the long term. This approach promotes the concept of increased harmony between government policy, land management practices and environmental constraints, leading to more sustainable agricultural production.

This chapter documents some of the recent progress that has been made in the United States, South Africa and Australia in drought mitigation. In each case, this progress is the direct result of a fundamental philosophical change by government. The development of drought policies that promote risk management rather than crisis management and the preparation of contingency plans represent a proactive step toward risk minimization and vulnerability reduction. Drought contingency plans promote greater coordination within and between levels of government, improve procedures for monitoring, assessing and responding to severe water shortages, and facilitate more efficient utilization of natural, financial and human resources.

REFERENCES

Anonymous. 1983. LGPA submits priorities for government assistance in future drought situations. *Livestock and Grain Producer* 6:1–3.

Australian Agricultural Council. 1983. An evaluation of existing drought policies given the current drought experience. Report by Standing Committee on Agriculture Working Group, Canberra, Australia.

Bruwer, J. J. 1989. Drought Policy in the Republic of South Africa. Proceedings of the SARCCUS Workshop on Drought, Alpha Training Center, June 1989.

Department of Primary Industry. 1984. *Review of the Natural Disaster Relief Arrangements*. Prepared for the National Drought Consultative Committee, Canberra, Australia.

Drought Policy Review Task Force. 1990. *National Drought Policy Final Report*. Canberra: Australian Government Publishing Service.

General Accounting Office. 1979. *Federal Response to the 1976–77 Drought: What Should Be Done Next?* Report to the Comptroller General, Washington, DC.

Glantz, M. H. 1987. Drought and economic development in sub-Saharan Africa. In *Planning for Drought: Toward a Reduction of Societal Vulnerability*, ed. D. A. Wilhite and W. E. Easterling. Boulder: Westview Press.

Great Lakes Commission. 1990. *A Guidebook to Drought Planning, Management and Water Level Changes in the Great Lakes*. Great Lakes Commission.

Interstate Conference on Water Policy. 1987. *Statement of Policy 1986–87*. Interstate Conference on Water Policy, Washington, DC.

Keating, P. J. 1984. Payments to or for the states, the Northern Territory and local government authorities, 1984–5. Treasurer of the Commonwealth of Australia, 1984–5. *Budget Paper No. 7*. Canberra, Australia.

National Academy of Sciences. 1986. *The National Climate Program: Early Achievements and Future Directions*. Washington, DC.

National Drought Consultative Committee. 1984. Drought assistance: financial arrangements. Notes from Meeting, March 28, 1984. Canberra, Australia.

National Farmers' Federation. 1982. *Drought Policy*. Canberra, Australia: National Farmers' Federation.

Orville, H. D. 1990. AMS Statement on Meteorological Drought. *Bulletin of the American Meteorological Society* 71:1021–3.

Pessoa, D. M. 1987. Drought in Northeast Brazil: impact and government response. In *Planning for Drought: Toward a Reduction of Societal Vulnerability*, ed. D. A. Wilhite and W. E. Easterling. Boulder: Westview Press.

Riebsame, W. E., Changnon, S. A. Jr and Karl, T. R. 1990. *Drought and Natural Resources Management in the United States: Impacts and Implications of the 1987–89 Drought*. Boulder: Westview Press.

Roux, P. W. 1991. South Africa devises a scheme to evaluate drought intensity. *Drought Network News* 3(3):18–23. Lincoln, Nebraska: International Drought Information Center, University of Nebraska.

Sinha, S. K., Kailasanathan, K. and Vasistha, A. K. 1987. Drought management in India: steps toward eliminating famines. In *Planning for Drought: Toward a Reduction of Societal Vulnerability*, ed. D. A. Wilhite and W. E. Easterling. Boulder: Westview Press.

Smith, J. B. and D. Tirpak (eds.) 1989. *The Potential Effects of Global Climate Change on the United States*. Environmental Protection Agency (EPA-230–05–89–050).

South Australian Department of Agriculture. 1983. Rural adjustment: interim report on drought relief measures. Submission to Industries Assistance Commission Inquiry. South Australian Treasury Department, Adelaide, Australia.

Stott, K. J. 1983. An economic assessment of assistance measures for the 1982–3 drought and for future droughts. *Internal Report Series*. Department of Agriculture, Victoria, Australia.

Thiruvengadachari, S. 1991. Satellite surveillance system for monitoring agricultural conditions in India. Paper presented at the Drought Management and Preparedness Training Seminar, Bangkok, Thailand, March. Sponsored by International Drought Information Center, University of Nebraska, Lincoln, Nebraska, UNEP and NOAA.

Western Governors' Policy Office (WESTPO). 1978. *Managing Resource Scarcity: Lessons from the Mid-Seventies Drought*. Institute for Policy Research.

Western States Water Council. 1987. *A Model for Western State Drought Response and Planning*. Western States Water Council.

WESTPO. 1977. *Director of Federal Drought Assistance, 1977*. Prepared by the Institute for Policy Research, WESTPO, for the Western Region Drought Action Task Force. Washington, DC: USDA.

Wilhite, D.A. 1986. Drought policy in the US and Australia: a comparative analysis. *Water Resources Bulletin* 22:425–38.

Wilhite, D. A. 1987. The role of government in planning for drought: where do we go from here? In *Planning for Drought: Toward a Reduction of Societal Vulnerability*, ed. D. A. Wilhite and W. E. Easterling. Boulder: Westview Press.

Wilhite, D. A. 1990. *Planning for Drought: A Process for State Government*. IDIC Technical Report Series 90–1.

Wilhite, D. A. 1991a. Drought planning and state government: current status. *Bulletin of the American Meterological Society* 72:1531–6.

Wilhite, D. A. 1991b. Drought planning: a process for state government. *Water Resources Bulletin* 27(1):29–38.

Wilhite, D. A. and Easterling, W. E. (eds.) 1987. *Planning for Drought: Toward a Reduction of Societal Vulnerability*. Boulder: Westview Press.

Wilhite, D. A. and Easterling, W. E. 1991. *Drought Management and Preparedness Training Seminar for Asia and Pacific Regions*. Final Report, March. Lincoln, Nebraska: International Drought Information Center, University of Nebraska.

Wilhite, D. A. and Glantz, M. H. 1985. Understanding the drought phenomenon: the role of definitions. *Water International* 10:111–20.

Wilhite, D. A. and Rhodes, S. L. 1994. State drought contingency planning: factors influencing plan development. *Water International* 19:15–24.

Wilhite, D. A., Rosenberg, N. J. and Glantz, M. H. 1986. Improving federal response to drought. *Journal of Climate and Applied Meteorology*. **25**:332–42.

World Meteorological Organization. 1986. *Model Drought Response Plan*. Memo to Permanent Representatives of Members of WMO, 14 May. Geneva, Switzerland.

IV

The International Conference on the Impacts of
Climatic Variations and Sustainable Development in
Semi-arid Regions (ICID)

9: Declaration of Fortaleza

International Conference on Impacts of Climatic Variations and Sustainable Development in Semi-arid Regions (ICID)

Fortaleza, Ceará, Brazil, 27 January to 1 February 1992

A contribution to the United Nations Conference on the Environment and Development (UNCED)

Sponsorship

Government of the State of Ceará
in conjunction with
Ceará State Federation of Industries (FIEC-CNI)
Bank of the Northeast of Brazil (BNB)
Esquel Group Foundation, Brazil (FGEB)

Organization

Esquel Group Foundation, Brazil

Support

National Research Council of Brazil (CNPq)
Interamerican Development Bank (IDB)
Institut Français de Recherche Scientifique pour le
Développement en Cooperation (ORSTOM)
The Government of the Netherlands
John D. and Catherine T. MacArthur Foundation
United Nations Development Program (UNDP)
United Nations Environment Program (UNEP)
World Bank (IBRD)
Brazilian Institute for the Environment and Renewable
Natural Resources (IBAMA)
Federal University of Ceará
Secretariat for Science and Technology, Presidency of the
Republic (SCT)
Brazilian Agricultural Research Company (EMBRAPA),
Ministry of Agriculture

J. Macedo Group
World Meteorological Organization (WMO)
United Nations Sudano-Sahelian Office (UNSO)

Site

Convention Center Edson Queiroz, Fortaleza, Ceará, Brazil

ICID

President of the Conference
Ciro Ferreira Gomes, former Governor of Ceará State

President of the International Advisory Committee
Tasso Ribeiro Jereissati, Governor of Ceará State

President of the National Advisory Committee
Fábio Feldman, Federal Deputy, President of the Environment Commission

President of the Organizing Committee
Antonio Rocha Magalhães, Director of ICID

Presentation

The International Conference on Impacts of Climatic Variations and Sustainable Development in Semi-arid Regions (ICID) brought together 800 participants including scientists, technicians, representatives of international and national public and private sector organizations, decision makers, and politicians from 45 different countries in the world. The event became an important opportunity for the exchange of experiences and information among many countries and regions.

During the Conference 63 papers were presented throughout 17 panel sessions, while 10 working groups met to carry out specific discussions. The outputs of the event include a political document – the Declaration of Fortaleza – approved in plenary session, and a summary of the discussions and recommendations of the 10 working groups. Both documents were delivered to the UNCED Secretariat as an attempt to include relevant issues concerning semi-arid regions in the UNCED agenda.

This chapter presents the Declaration of Fortaleza.

THE DECLARATION OF FORTALEZA ——

Whereas:

many semi-arid regions are economically marginal, therefore highly vulnerable to any changes in the global climate, and often lack the financial and technical resources required to initiate major adaptations to environmental changes;
there is a growing recognition of the need for international action on the issues of global change, environmental degradation and equity; and
there is great need for urgency arising out of current conditions in the semi-arid regions of dveloping areas:

1. We, women and men of the civil society, have met in Fortaleza, State of Ceará, Brazil, from 27 January to 1 February 1992, convened by the Government of the State of Ceará and Fundação Grupo Esquel Brasil, with the support of several other national and international organizations. We are a multidisciplinary group of scientists, academics, government officers, social workers, environmentalists, politicians, religious leaders and other persons concerned with the interactions between human beings and their natural and social environment, and have many decades of study, research and practical experience in the development of semi-arid regions worldwide.

2. We address all women and men concerned with equitable development and a decent life for everyone in harmony with nature, and also the women and men empowered by society to take the crucial decisions related to the welfare of present and future generations.

3. We issue this Declaration at a particularly important moment, since in June 1992 the United Nations Conference on the Environment and Development (UNCED) will meet in Rio de Janeiro, Brazil. The determination, energies and resources of all countries of the world will be mobilized in a great effort to stop the processes of environmental deterioration that threaten our future. It would be a great loss to humanity if, on this occasion, the plight of semi-arid

regions and their environmental degradation are not recognized and discussed by the nations attending UNCED.

4. Although semi-arid zones in all regions of the world are at risk, our primary concern is the present and future conditions of people and their environment in the semi-arid regions of developing countries. These regions are characterized by their extreme vulnerability to highly variable rainfall regimes and inadequate human activities, and they are places where great human suffering occurs: abject poverty, recurring famines and outmigration, uncertainty of agricultural production from one year to the next, and consequently uncertainty about the continuity of human settlements and their rich cultures and civilizations. Though semi-arid regions have common problems, they have not sought sufficiently to strengthen each other by sharing experiences and knowledge.

5. The deterioration of the environmental and human conditions in these regions, which in many cases include significant desertification processes, has widespread socioeconomic consequences which affect directly and indirectly all the regions of the world. In addition, as semi-arid ecosystems are fragile they are highly susceptible to effects of global warming and other environmental changes brought about by unsound industrial and agricultural activities and unwise, unsustainable development practices.

6. The specific recommendations widely discussed and supported by the participants in ICID may be summarized in the following basic principles, which should guide efforts to develop semi-arid regions:

    i.   The origin of poverty and environmental degradation in these areas is basically socioeconomic and political. Any serious effort to develop those regions should be based on socioeconomic strategies that include the participation of civil society and political commitment, structural reforms, land-tenure reform, access to water adequately managed and improved agricultural policies at national and regional levels. The overall improvement of education and skills is a precondition for the achievement of the development goals of these strategies.

    ii. The environmentally sustainable economic and social development of semi-arid regions must be pursued as the ultimate goal. Sustainable development must be understood as including equitable distribution of wealth and access to resources, respect for local diversity, an adaptation of the scale of human activities to those which are compatible with the ecological regime taking into account the needs of present and future generations, and maintenance and increase of long term productivity.

    iii. Past and present mistakes and ill-conceived policies have produced a situation in those regions where most of them have reached the limits of their productive capacity, a situation which can not be reversed without large national and international financial efforts. A fair approach to this problem indicates that some costs will have to be borne by the national economies but that others are clearly the responsibility of the international community. Efforts to define the scope of each one's financial responsibility must be associated with the increase in the capacity of developing countries to finance their own sustainable development, including the appropriate resolution of the external debt burden of indebted developing countries. Trade discrimination particularly hurts the semi-arid regions, and should be removed. It is also essential that semi-arid regions be assured access to new technologies, and international rules to promote the transfer of technologies should be implemented as soon as possible. In addition, it is emphasized that measures aimed at the restoration of degraded areas and prevention of further environmental deterioration are urgently needed.

iv. Efforts to redress the present impoverished situation must be based on rigorous research and study but can not neglect the knowledge of the traditional populations, who for centuries sustained life in those regions. These efforts will not reach their objectives if civil society and 'grass roots' movements are not involved in the national and regional decision making process.

v. Biodiversity has an actual and potential economic importance. The development of these regions should take into account the sustainable use and conservation of this resource. Adequate national legislation and international conventions should prevent abusive exploitation. The countries, cultures and regions from which each particular genetic resource or its natural or synthetic derivates originate must be fairly compensated by those who acquire or transform them for economic gain.

vi. Semi-arid regions must learn from each other. Networks should be created and existing ones strengthened. They would serve two needs: at the research level they would be used to exchange information, discuss methodologies, report research findings and develop joint activities; at the level of planning for sustainable development, they will provide a forum for dialogue and exchange of experience among experts, decision makers and organized segments of civil society.

vii. History shows, and modern science confirms, that semi-arid regions have the natural resources required for the development of human settlements where it is possible to establish adequate production systems, equitable forms of social organization and rich and prosperous cultures. We are not facing hopeless situations but a set of environmental and socioeconomic circumstances which require special attention and priority treatment to allow for the full development of the potentialities of semi-arid regions and their peoples.

10: Highlights of Working Group Discussions and Recommendations

This chapter summarizes the discussions and recommendations of 10 Working Groups convened as part of the International Conference on the Impacts of Climatic Variations and Sustainable Development in Semi-arid Regions (ICID) in Fortaleza, Brazil, 27 January to 1 February 1992.

CRITICAL ISSUES AND NEEDS OF SEMI-ARID REGIONS

Climate variability and climate change

Semi-arid regions exist in the tropical, sub tropical and temperate zones of the earth. Their most prominent common characteristic is a lack of sufficient, reliable and timely rainfall. They experience wide variations in seasonal and annual precipitation. Droughts occur periodically. In recent years, scientists have made encouraging progress in predicting seasonal to inter-annual variability of precipitation in these critical zones, particularly in the tropics. At other times, extreme climatic variability may result in flooding. Under past and current conditions semi-arid regions have been climatic risk zones. A number of factors compound this natural vulnerability.

An increase in population and more intensive human activities, as has been experienced in many semi-arid regions, leads to increased stress on fragile or scarce resources, such as water, soil and wildlife. In many areas the population is in excess of the ecological carrying capacity under the current systems of economic production and resource use.

Many semi-arid regions are in developing countries and suffer from a high incidence of poverty, underemployment, poor health and illiteracy. Public infrastructure is often deficient, including roads, water resources, industry and housing. Economically and politically, rural populations often are powerless. The combination of climatic and social stress increases the vulnerability of the natural and human systems of semi-arid regions.

The regional consequences of global warming cannot yet be predicted with precision and confidence. But some impacts are probable. Increases in temperature will lead to increased evapotranspiration. This will be an important factor in places where the climate is hot under current conditions. Whether rainfall in these regions will increase or decrease is not known. The International Panel on Climate Change (IPCC) lists semi-arid regions among those areas likely to experience increased climatic stress. Semi-arid regions commonly exhibit steep gradients of precipitation. The spatial shifts of these zones should be examined now, because adverse effects may occur long before man-made climate change can be reliably detected. It is also possible that climatic change will have unexpected consequences for semi-arid regions.

The problems that climate poses today may become more frequent and more severe with global warming. Research on and responses to climate variability and climate change share many common features. Understanding how societies, in the past and now, cope with climate variability in semi-arid regions can provide a first approximation of how to prepare for climate change in future decades.

Social impacts

Climate–society interactions need to be better understood. This is particularly important for regions with extreme climates. In the case of the semi-arid regions we must identify those human activities that increase vulnerability to climatic variability and change. Historically, small populations have maintained subsistence agriculture in semi-arid regions over long periods of time. But they cannot provide sustainable livelihoods because they are highly vulnerable to extreme climatic events and cannot support large populations. Population growth, as well as out-migration during droughts, will increase the vulnerability of natural and social systems in semi-arid regions as well as in the areas that surround them. The exodus of people from these marginal areas to shanty towns, urban slums and to the forests is vivid proof of the lack of sustainability. This cycle of poverty is maintained as people do not gain enough education and training to improve their condition permanently.

Economic impacts

Semi-arid regions produce subsistence crops, as well as raw materials that are sought after by other nations. There is little value added in the region, due to a lack of expertise and local firms. Economic power is often concentrated among a few large landholders or companies with headquarters elsewhere. The people living in the region often do not own land, may have limited access to water, and lack access to knowledge and technology that would allow them to work the land using improved practices. At the same time, semi-arid regions face economic disadvantages: transportation and production costs (fertilizer, pesticides) are high due to their remote locations, cold storage is lacking, crop choice is limited, and access to export markets is difficult or non-existent. For some commodities, protectionist trade barriers of importing countries restrict the export potential of semi-arid regions.

Investment for development of infrastructure (transportation, communication, water, electricity) is often insufficient. In many cases, national governments lack the necessary resources, especially those of developing countries with large foreign debts.

These characteristics make the economy of semi-arid regions highly vulnerable to climate variations. Frequent droughts affect mainly the agricultural sector, causing heavy production and productivity losses. Consequently, agricultural employment drops sharply during droughts. Subsistence agriculture and small farms are affected more than cash crops and large landholdings, including livestock, but these are also greatly affected by extreme climatic events.

Environmental impacts

Natural ecosystems in semi-arid regions are fragile. Each degraded component of the ecosystem also degrades other parts of the environment. Desertification and other forms of land degradation, resulting from natural or human causes, may be an irreversible process leaving behind land that is permanently lost to productive use. Many factors contribute to land degradation and must be considered in efforts to stop further degradation: soil erosion, sedimentation, salinization, waterlogging, deforestation, overgrazing, soil compaction and crusting, unsuitable patterns of land tenure, inappropriate use of water resources, urbanization and poor land management. Each of these factors is highly sensitive to climatic variations, and most probably to climate change. Strategies to deal with land degradation exist, requiring the integration of research, technology and policy. Failure on this front will preclude the attainment of sustainable devel-

opment in semi-arid regions. The key to success is a much improved and timely integration of knowledge with action.

Sustainable adaptation strategies

Human adaptation to semi-arid conditions provides useful lessons for how societies might cope with climate-related environmental changes in the future. The most general lesson, perhaps, is that the success or failure of human responses is determined by the interplay between climatic, socioeconomic and political factors. A second lesson is that the experience of people who have long lived under semi-arid conditions needs to be understood, evaluated and used in developing appropriate response strategies. This has rarely been done in a systematic and sustained fashion. Another important lesson is that researchers from outside the region must be careful not to impose inappropriate research frameworks. This suggests a greater need for local participation in research, planning and implementation of policy. The old paradigm of development by heavy capital investment and following the path of 'northern' industrialization is not sustainable in the long term. The new paradigm for sustainable development, while still largely a goal toward which to aim, must be adopted.

Key components of sustainable development include a new system of social and economic accounting, increased use of appropriate technologies, a more equitable distribution of income and quality of life, increased political participation and decentralization of planning and decision making. While sustainable development is difficult to implement in any ecological setting, this is even more the case in resource-stressed and poor environments. Yet though the difficulties are larger, the potential benefits are important. Sustainable development in semi-arid regions may provide the only answer to survival.

The example of land reform illustrates the special need for sustainable development in semi-arid regions. In many places it is not so much the absolute lack of water or food but access to and distribution of available resources that is the problem. Without equitable access to water, food and land the region, and all of its people, will not have a sustainable future. A method to implement land reform, and other actions to create a more favorable environment for sustainable development, may involve a new system of negotiating and bargaining between social, political and economic interests. Solutions will be difficult to reach unless all partners to the bargain perceive an advantage. Land reform, using such a social consensus, may provide the opportunity for the landless to own land and to move from subsistence to production agriculture, while the large landowners may focus on value-added production and thereby create improved conditions for exporting their goods.

Linkage to atmospheric pollution

Semi-arid regions, on the whole, do not contribute large amounts of pollutants to the atmosphere. Even so, increasing population and industrialization in these regions leads to emissions of conventional pollutants as well as greenhouse gases. Biomass burning, cattle raising, the burning of fossil fuels, industrial production, mining and urbanization are the principal causes. Such pollution poses risks to human health, and may change local and global climates. Development plans for semi-arid regions must consider these effects.

Financing sustainable development

Neither national governments nor international organizations have to date given high priority to investments in semi-arid regions. The funding that was received has often been poorly used and has done little to improve social, economic or environmental conditions in the recipient regions. Increased investment is needed, and it should be provided in a timely fashion. Semi-arid regions, in order to avoid further degradation, must receive increased support from international agencies and from their national governments. It is equally important that funding be made available in a sustained and predictable fashion.

Funding criteria need to be redefined in order to meet the objectives of sustainable development. This requires a better assessment of all costs and benefits involved in new programs. Environmental and ecological costs need to be explicitly included. Ecologically sound investments must be financed, even though they will be more expensive than traditional programs. Such investments are entirely justified as long as they are cost effective.

Multinational development agencies should create specialized funding mechanisms for the needs of semi-arid regions, which would take timely action in response to requests by national governments. The system for identifying, designing, appraising, implementing and monitoring for international financing of projects must become more efficient and timely, and must provide opportunities for more local participation and feedback.

Research needs

Studies of societal impacts of and response strategies to extreme climatic events in semi-arid regions (droughts, storms, floods, frosts, etc.) are needed. The process of policy making and institutional development in semi-arid regions should be addressed. Water and agricultural policies and management practices should receive particular attention. Emissions of pollutants and greenhouse gases originating from these regions need to be studied. The effectiveness of ongoing national and international research and assistance programs for semi-arid regions must be evaluated. Agricultural, biological and genetic research addressed to the specific needs of semi-arid regions should receive high priority from sponsoring agencies. Finally, a better understanding of society–climate interaction is needed. This is the basis for changing behavior, and is not well understood.

Networking and cooperation

Semi-arid regions around the globe share many common problems despite differences in local conditions. They can learn from each other and can develop joint activities. For this purpose, we recommend that networks be created and existing ones strengthened to address problems that are common to them. This would serve at least two needs: At the research level, experts from different disciplines could exchange information, discuss methodologies, report research results, and develop joint activities; at the level of planning for sustainable development, they could provide a forum for dialogue and exchange of experience among experts and decision makers. They could also provide opportunities for education, training and exchanges. Networks may be organized internationally or among those regions having similar characteristics. One possibility would be to create a multinational network for semi-arid regions with specialized components for tropical, sub-tropical and temperate zones.

Networks should meet the following criteria:

- participation should be based on a bottom-up strategy, with strong participation by local and regional organizations;
- a wide range of public and private organizations should participate, including farmers, women, labor, business and environmental groups;
- South–South networking should receive priority attention;
- maximum use should be made of existing mechanisms and new organizations created only if existing ones cannot perform the necessary tasks;
- one-to-one partnerships should be created between semi-arid regions in the more prosperous countries and developing countries.

Within each semi-arid region a concerted effort must be made to plan for a sustainable future. Measures to cope with the problems of semi-arid regions need to be coordinated by organizations and people within the regions. This requires cooperation among experts from different fields of research, government agencies and the civil society.

A consortium or roundtable, operating within a semi-arid region, should be convened and staffed by a government agency, a university or a non-governmental organization. It is important that such an effort be sustained over time, so that studies can be undertaken and the results used for planning and implementation of sustainable development practices. Each semi-arid region should develop plans for such a consortium to be convened at the earliest possible time. Financial support for these consortia could be provided, for example, by national and provincial governments, as well as regional development agencies. In the case of developing countries, additional support should be provided by international and regional financial institutions.

High-level policy discussions should focus on the problems and prospects of semi-arid regions. For this, the United Nations should convene a standing committee. This group could be made up of government officials, experts and representatives of non-governmental organizations. It would be charged with two primary tasks: (1) to assist with the development of an international action program for semi-arid regions, consulting with a wide range of organizations about its goals; and (2) to give support to development plans prepared by the individual semi-arid regions.

SUMMARY OF RECOMMENDATIONS

Within a context of limited resources, widespread poverty, environmental degradation and changing climate, the basic challenge in the semi-arid regions is to find a path for sustainable development that guarantees to their populations that basic material needs are meet, income is distributed more equitably, and natural resources of these regions are preserved.

To achieve this goal a concerted effort is necessary at national and international levels, along the following lines of action:

I. Improvement of the capacity of people, particularly in rural areas, to influence and control their future, through the ownership or usufruct of assets, broadly defined to include also human assets. Investments in education, training, health, sanitation and research, on the one hand, and agrarian reform and financing of productive activities, on the other, are the most relevant segments of this strategy.

II. Empowerment of the population, through greater participation and decentralization in the planning and decision-making process. Planning agencies should adopt procedures for integrated impact assessments in order to identify problems, collect data, and provide opportunities for public discussion and participation. Projects should be conceived and developed using a bottom up, rather than a top-down approach.

III. Mobilization of resources from internal and external sources (bilateral and multilateral) at levels suffi-

ciently high to generate an impact in terms of alleviation of poverty in semi-arid regions.

A. Multilateral development agencies should create a specialized financial and technical facility designed to respond to the specific needs of semi-arid zones which would react effectively and in timely fashion to concepts and initiatives presented by national governments.

IV. Improvement in the efficiency of the management of natural resources and in the use of capital funds, existing or newly mobilized.

A. Limited resources in semi-arid regions should be assessed and managed with special care. Production processes designed for other regions are not always appropriate.

B. The management strategy should combine grass roots knowledge with current scientific and technological contributions. Policies should include both formal and informal education, which involves learning from and advising and supporting local groups.

C. Government in semi-arid regions should establish 'extension services' to educate practitioners in agriculture, industry and other activities about new available technologies and practices.

D. All interventions must be proven to be sustainable in terms of use and availability of natural resources and must be ecologically sound.

E. A radically different approach to the identification, design, appraisal, implementation, monitoring and follow-up of projects shall be established, focusing on:

- efficient, flexible and timely response to needs identified in conjunction with participating communities;

- procedures which provide effective monitoring of interventions allowing for feedback and appropriate adjustment as required;

- decentralization of decision making and implementation which incorporates to the maximum extent the capabilities of the communities affected as well as non-governmental organizations.

F. In project identification and design, a more precise definition of the intended target groups as well as their own participation is required.

G. Continued integrated monitoring, observation and multi-disciplinary analysis of historical and contemporary responses to climate variability should be undertaken to determine which sub-groups or sectors of society are most sensitive to climatic variation. Improved awareness of the uniqueness and fragility of those areas is critical for a strategy of sustainable development.

V. Lack of scientific information on natural and socio-economic systems in semi-arid regions is considered one of the most serious threats to sustainable development. Therefore a comprehensive and intensive research program should be developed along the following agenda:

A. The origins, consequences of and response strategies to the extreme phenomena related to climatic variability in semi-arid regions (e.g. droughts, storms, floods, frosts).

B. Effectiveness of current national and international cooperation programs in semi-arid regions, to provide reliable identification of the deficiencies existing today and to improve information gathering, retrieval and exchange as well as technology transfer and transfer of research facilities and personnel among industrialized and developing nations. Emphasis should be given to comparative studies with special reference to the damage provoked by climatic variations, and risks they pose to particular countries.

C. Policy-making and management systems.

VI. To understand the complex interactions between environmental and human systems the following actions are proposed:

A. Integrated regional impact assessments are required. These should be collaborative efforts involving regional experts and international research and funding organizations. Cooperative research institutes should be established in specified semi-arid regions with long-term support of international organizations.

B. To improve knowledge of air quality and emissions in semi-arid regions, a permanent network of monitoring stations should be established and integrated with the WMO-GAW (Global Atmosphere Watch) network.

C. Emissions from the industrial, energy and transportation sectors must be reduced and controlled through the establishment of air quality standards.

D. There should be additional international and national investment in the development of efficient decentralized small energy production systems for rural semi-arid regions, with specific emphasis on renewable energy technologies.

To implement this program the following actions are most urgently needed: funding for research, cooperation of national institutions at the international level, technology transfer and education.

VII. Cooperation among and within semi-arid regions should be improved. The following actions are recommended:

A. Organization of networks, internationally or between regions with similar characteristics, for joint research and exchange of experience among experts and decision-makers.

B. Creation of consortia or roundtables within semi-arid regions to plan and coordinate actions for sustainable development. The consortia would integrate the efforts of governmental agencies, universities, research centers and non-governmental organizations.

C. Organization of a standing committee at the United Nations to focus high-level discussion on the problems and prospects of semi-arid regions.

VIII. The issue of population growth in semi-arid regions should be addressed taking into account:

A. An agro-ecological mapping to identify areas where pressure upon natural resources poses a threat to the survival of the population.

B. The need to plan the development of those areas by harmonizing the economic interventions with the natural resource base and by combining productive and social investments, particularly environmental education.

C. The urgency to alleviate poverty through increased investments and employment, financed by income transfer from more developed areas, at the national and international levels.

D. The search for local solutions of economic and technological problems since transfers of populations can result only in the transfer of pressures to other environments.

COMMITTEES

International Committee

Taso Ribeiro Jereissati, Ceará State Governor
Albano Franco, President of CNI
Alfredo Costa Filho, CEPAL/ILPES
Donald Wilhite, IDIC, University of Nebraska
George Woodwell, Woods Hole Research Center, MA
Howard Ferguson, SWCC, Geneva
Jean-Pierre Raison, Université Nanterre-X
Jorge Hardoy, Esquel Group Argentina
Jorge Lins Freire, President of the Bank of Northeast of Brazil
Juan Felipe Yriart, President of Esquel Group Foundation
Judith Tendler, MIT, Cambridge, MA
Jurgen Schmandt, HARC, The Woodlands, TX
Kilaparti Ramakrisna, Woods Hole Research Center, MA
Les Heathcote, Flinders University, Australia
Martin Parry, AIR Group, Oxford University, UK
Michael Glantz, NCAR/ESIG, Boulder, CO
Nancy Birdsall, World Bank, Washington, DC
Norman Rosenberg, Resources for the Future, Washington, DC
N. S. Jodha, ICIMOD, Kathmandu, Nepal
Peter Usher, UNEP, Nairobi, Kenya
Phyllis Pomerantz, World Bank, Washington, DC
Roberto Mizrahi, President of Esquel Group, New York, NY
Severino de Melo Araujo, FAO, Santiago, Chile
Stahis S. Panagides, Esquel Group, Washington, DC
Thomas Downing, AIR Group, Oxford, UK
William Riebsame, Natural Hazards Center, Boulder, CO

National Committee

Fábio Feldman, Federal Deputy, Brasília
Antônio Albuquerque de Sousa Filho, Rector, Federal University of Ceará
Antônio Carlos do Prado, IBAMA, Brasília
Antônio Divino Moura, INPE, São Paulo
Augusto Pires, University of Brasília
Carlos Afonso Nobre, INPE, São Paulo

Enéas Salati, President of INPA, Manaus
Eustáquio Reis, IPEA, Rio de Janeiro
Flora Cerqueira, UNDP, Brasília
Gustavo Maia Gomes, Federal University of Pernambuco
Hélio Barros, Secretariat of Science and Technology – PR, Brasília
Jorge Santana, SUDENE, Recife
José Vieira do Nascimento, CNI, Rio de Janeiro
Manuel Tourinho, EMBRAPA, Brasília
Mauro Benevides Filho, Secretary of Planning, Fortaleza
Nilson Holanda, University of Brasília
Osmundo E. Rebouças, former Federal Deputy
Oswaldo Massambani, University of São Paulo
Rubens Vaz da Costa, former Secretary of Energy of Brazil

Organizing Committee

Antônio Rocha Magalhães, Esquel Group, Brasília
Marfisa Aguiar, Secretary of Urban Development and Environment, Fortaleza
Adolfo de Marinho Pontes, Secretary of Social Action, Fortaleza
Agostinho Fernandes Bezerra, Esquel Group, Brasília
Eduardo de Castro Bezerra Neto, UECE, Historical Institute, Fortaleza
Benito Moreira de Azevedo, ICID, Fortaleza
Elizabeth Machado Duarte, Esquel Group, Brasília
Adriana Moura, Esquel Group, Brasília
Francisco José da Silveira, Esquel Group, Brasília
Faustino de Albuquerque Sobrinho, UFC, Fortaleza
Francisco Lopes Viana, FUNCEME, Fortaleza
Hermano Frank, FIEC, Fortaleza
Lincoln Coutinho de Aguiar, BNB-ETENE, Fortaleza
Luiz Carlos Tavares, CNPq, Brasília
Luiz Esteves Neto, President of FIEC, Fortaleza
João Fontenele, FIEC, Fortaleza
Jessé Cláudio Fontes de Alencar, CNI, Rio de Janeiro
Marie-Madeleine Mailleux Sant'Ana, Esquel Group, Brasília
Paula Dias Pini, Esquel Group, São Paulo
Pedro Albuquerque, Equatorial Institute, Fortaleza
Sílvio Rocha Sant'Ana, Esquel Group, Brasília